Semiconductor Physics Modeling Nanostructured Semiconductors

Jamie Flux

https://www.linkedin.com/company/golden-dawn-engineering/

Contents

1 Introduction to Nanostructured Semiconductors   **11**
    Nanostructured Materials 11
    1   Quantum Wells 11
    2   Quantum Wires 12
    3   Quantum Dots 12
    4   Nanostructured Heterojunctions 12
    Challenges in Modeling Nanostructured Semiconductors 12
    1   Size and Shape Effects 13
    2   Heterostructure Interface Quality 13
    3   Computational Complexity 13
    4   Multi-Physics Coupling 13
    5   Experimental Validation 13
    Python Code Snippet 14

2 Quantum Wells   **17**
    Introduction to Quantum Wells 17
    Modeling Quantum Well States 18
    Solving the Schrödinger Equation 18
    Electronic and Optical Properties 19
    Conclusion 19
    Python Code Snippet 19

3 Quantum Wires   **22**
    Introduction to Quantum Wires 22
    1   One-Dimensional Schrödinger Equation . . . 22
    2   Quantum Wire Potential 23
    Solving the Schrödinger Equation 23
    1   Boundary Conditions 23
    2   Solutions for Quantum Wire States 23

	Transport and Confinement Effects	24
1	Quantum Confinement	24
2	Wavefunction Localization	24
3	Quantum Wire Conductance	24
	Conclusion	24
	Python Code Snippet	25

4 Quantum Dots — 27

	Introduction to Quantum Dots	27
1	Size and Shape Effects	27
2	Excitonic Properties	28
	Modeling Quantum Dots	28
1	Effective Mass Approximation	28
2	Poisson-Schrödinger Equation	28
3	Density Functional Theory (DFT)	29
4	Configuration Interaction (CI)	29
	Conclusion	29
	Python Code Snippet	29

5 Nanostructured Heterojunctions — 33

	Introduction to Nanostructured Heterojunctions	33
1	Band Alignment at Interfaces	33
2	Charge Transport in Heterojunctions	34
	Mathematical Modeling of Heterojunction Interfaces	34
1	The Drift-Diffusion Model	34
2	Schrodinger-Poisson Solver	35
3	Kinetic Monte Carlo Simulations	35
4	Advanced Approaches	35
	Conclusion	35
	Python Code Snippet	36

6 Nano-MOSFETs — 39

	Introduction to Nano-MOSFETs	39
1	MOSFET Scaling	39
2	Quantum Mechanical Effects	40
	Mathematical Modeling of Nano-MOSFETs	40
1	Drift-Diffusion Model	40
2	The Effective Mass Approximation	41
3	Non-Equilibrium Green's Functions	41
4	Monte Carlo Simulations	41
	Conclusion	41
	Python Code Snippet	42

7 Band Structure Engineering — 46
Introduction to Band Structure Engineering 46
1   Methods for Band Structure Engineering .. 47
Computational Methods for Band Structure Engineering 48
1   Tight-Binding Method 48
2   Density Functional Theory 48
3   Empirical Pseudopotential Method 49
4   Multi-Scale Modeling Approaches 49
Conclusion 49
Python Code Snippet 50

8 Nanostructured Photonic Crystals — 53
Introduction 53
Fundamentals of Photonic Crystals 53
Modeling Light-Matter Interactions 54
1   Maxwell's Equations 54
2   Finite Difference Time Domain (FDTD) Method 54
Defect Modes in Photonic Crystals 55
1   Plane Wave Expansion Method 55
Conclusion 55
Python Code Snippet 56

9 Thermoelectric Nanostructures — 59
Introduction 59
Quantum Transport Modeling 60
1   Boltzmann Transport Equation 60
2   Density Functional Theory 60
Phonon Transport Modeling 61
1   Boltzmann Transport Equation for Phonons . 61
2   Molecular Dynamics Simulations 61
Coupled Transport Modeling 62
1   Coupled Electron-Phonon Transport Equations 62
Conclusion 62
Python Code Snippet 63

10 Plasmonic Nanostructures — 65
Introduction 65
Maxwell's Equations and Plasmonics 66
Finite-Difference Time-Domain (FDTD) Method .. 66
Boundary Element Method (BEM) 67

Conclusion	68
Python Code Snippet	69

11 Carrier Dynamics in Nanostructures   72
Introduction	72
Carrier Lifetimes	73
Carrier Mobility	73
Recombination Processes	74
Conclusion	74
Python Code Snippet	74

12 Atomistic Simulations   77
Introduction	77
Atomic Interactions	78
Numerical Algorithms	78
Applications	79
Conclusion	80
Python Code Snippet	80

13 Self-Assembly Processes   83
Introduction to Self-Assembly		83
1	Fundamental Interactions	84
Mathematical Modeling of Self-Assembly		84
1	Lattice Models	85
2	Continuum Models	85
Simulation Techniques		85
1	Molecular Dynamics (MD)	86
2	Monte Carlo (MC)	86
3	Brownian Dynamics (BD)	86
4	Phase Field Methods	86
Conclusion		87
Python Code Snippet		87

14 Strain Engineering in Nanostructures   90
Introduction to Strain Engineering		90
1	Strain and Elasticity	91
2	Effects of Strain in Nanostructures	91
Mathematical Modeling of Strain in Nanostructures		92
1	Elastic Continuum Models	92
2	Strain-Gradient Elasticity Models	93
3	Atomistic Simulations	93
Computational Techniques for Strain Engineering		94

1　Finite Element Method (FEM) 94
　　2　Boundary Element Method (BEM) 94
　　3　Molecular Dynamics (MD) 94
　　Conclusion 95
　　Python Code Snippet 95

15 Nanostructure Fabrication Techniques · 98
　　Introduction to Nanostructure Fabrication 98
　　1　Importance of Fabrication in Nanotechnology　98
　　2　Modeling Challenges in Nanostructure Fabrication 99
　　Mathematical Modeling of Nanostructure Fabrication　99
　　1　Kinetic Monte Carlo Simulations 99
　　2　Diffusion Models 99
　　3　Phase Field Models 100
　　4　Finite Element Analysis 100
　　Computational Techniques in Nanostructure Fabrication 100
　　1　Molecular Dynamics (MD) 101
　　2　Quantum Mechanical Methods 101
　　3　Computational Fluid Dynamics (CFD) 101
　　4　Data-driven Modeling 101
　　Conclusion 102
　　Python Code Snippet 102

16 Nanostructured Solar Cells · 106
　　Introduction to Nanostructured Solar Cells 106
　　1　Importance of Modeling Photovoltaic Mechanisms 106
　　Fundamentals of Nanostructured Solar Cells 107
　　1　Absorber Layer 107
　　2　Charge Transport Layers 107
　　3　Electrodes 107
　　Mathematical Modeling of Nanostructured Solar Cells 108
　　1　Optical Modeling 108
　　2　Charge Generation Modeling 108
　　3　Charge Transport Modeling 108
　　4　Device-Level Modeling 109
　　Conclusion 109
　　Python Code Snippet 109

17 Excitonic Effects in Nanostructures — 112
- Excitons: Basic Concepts and Properties — 112
  - 1 Exciton Formation and Binding Energy — 112
  - 2 Exciton Size and Spatial Distribution — 113
- Excitonic Effects on Nanomaterial Properties — 113
  - 1 Optical Absorption and Emission — 114
  - 2 Exciton Dynamics and Luminescence — 114
  - 3 Exciton-Exciton Interactions — 115
- Conclusion — 115
- Python Code Snippet — 115

18 Photoluminescence in Nanostructures — 119
- Principles of Photoluminescence — 119
  - 1 Rate Equations — 120
  - 2 Radiative and Non-Radiative Recombination — 120
- Mathematical Models for Photoluminescence — 121
  - 1 The Rate Equation Model — 121
  - 2 The Detailed Balance Model — 121
- Applications of Photoluminescence in Optoelectronic Devices — 121
  - 1 LEDs and Lasers — 122
  - 2 Photovoltaic Cells — 122
- Conclusion — 122
- Python Code Snippet — 122

19 Multiphysical Simulations — 125
- Introduction to Multiphysical Simulations — 125
  - 1 Motivation and Importance — 125
  - 2 Challenges and Considerations — 126
- Mathematical Formulation of Multiphysical Simulations — 126
  - 1 Governing Equations — 127
  - 2 Coupling Terms — 127
  - 3 Solution Techniques — 127
- Applications of Multiphysical Simulations — 128
  - 1 Microelectronics — 128
  - 2 Energy Systems — 128
  - 3 Biomechanics and Biomedical Engineering — 128
- Conclusion — 129
- Python Code Snippet — 129

20 Hybrid Nanostructures   132
Introduction to Hybrid Nanostructures 132
1   Characteristics of Hybrid Nanostructures . . 133
Modeling Approaches for Hybrid Nanostructures . . 133
1   Continuum Models 134
2   Particle-based Models 134
3   Multiscale Models 135
Applications of Hybrid Nanostructures 135
1   Optoelectronics 135
2   Sensing and Biosensing 136
3   Energy Conversion and Storage 136
4   Biomedical Devices and Therapeutics 136
Conclusion 137
Python Code Snippet 137

21 Graphene and 2D Materials   140
Introduction to Graphene and 2D Materials 140
1   Crystal Structure of Graphene 141
2   Electronic Band Structure of Graphene . . . 141
3   Other 2D Materials 141
Modeling Approaches for Graphene and 2D Materials 141
1   Electronic Band Structure Calculations . . . 141
2   Transport Modeling 142
3   Strain and Defect Engineering 142
4   Optical and Photonic Properties 142
5   Heterostructures and van der Waals Interactions 142
Conclusion 143
Python Code Snippet 143

22 Quantum Transport in Nanostructures   146
Theoretical Framework for Quantum Transport . . . 146
1   Mathematical Description of the Schrödinger Equation 147
2   Boundary Conditions for Nanostructures . . . 147
Transport Parameters and Quantum Mechanics . . . 147
1   Transmission Coefficient 147
2   Landauer-Büttiker Formula 148
3   Density of States 148
Computational Methods for Quantum Transport . . 148
1   Transfer Matrix Method 148
2   Green's Function Method 149

    3   Non-Equilibrium Green's Function Formalism 149
    Conclusion 149
    Python Code Snippet 150

23 Localized Surface Plasmon Resonance (LSPR)   153
    Introduction to Localized Surface Plasmon Resonance 153
    Mathematical Modeling of LSPR 154
    1   Maxwell's Equations for LSPR 154
    2   Boundary Conditions and Material Properties 154
    3   Plasmon Resonance Condition 155
    Applications of LSPR in Sensing 155
    1   Nanoparticle-based LSPR Sensors 155
    2   LSPR in Surface Enhanced Raman Spectroscopy
         (SERS) 155
    3   LSPR-based Biosensors 156
    Conclusion 156
    Python Code Snippet 156

24 Surface Chemistry in Nanostructures   159
    Surface Adsorption and Desorption Processes 159
    Surface Modifications and Functionalization 160
    Surface Interface Models 160
    Surface Catalysis and Reactivity 161
    Conclusion 162
    Python Code Snippet 162

25 Nanomagnetism   165
    Magnetic Moments and Spins 165
    Exchange Interactions 166
    Magnetic Anisotropy 166
    Dynamics of Nanomagnetism 167
    Technological Implications 167
    Conclusion 168
    Python Code Snippet 168

26 Spin Dynamics in Nanostructures   171
    Spin and Its Quantum Description 171
    Spin Dynamics in Magnetic Fields 172
    Spin-Orbit Interaction 172
    Spin Transport and Spin Currents 173
    Conclusion 173
    Python Code Snippet 174

27 Charge Trapping and Defects · · · · · · · · · · · 177
Introduction 177
Charge Trapping Mechanisms 177
  1   Trap States and Energy Levels 178
  2   Carrier Capture and Emission Processes . . . 178
  3   Trap Filling and Emptying Dynamics 178
Defects in Nanoscale Semiconductors 178
  1   Defect Energy Levels 179
  2   Defect-Induced Strain and Carrier Mobility . 179
  3   Impact on Device Performance 179
Conclusion 179
Python Code Snippet 180

28 Nanostructured Field-Effect Devices · · · · · · 182
Introduction 182
Fundamentals of Field-Effect Devices 182
  1   Basic Operation Principles 183
  2   Traditional FET Structures 183
Benefits of Nanostructuring in FET Technology . . . 183
  1   Improved Control Over Channel Conductance 183
  2   Enhanced Electrostatic Control 184
  3   Novel Device Architectures 184
  4   Integration with Other Nanomaterials 184
Conclusion 184
Python Code Snippet 185

29 Multi-Scale Modeling Approaches · · · · · · · 188
Introduction 188
Atomic Scale Modeling 188
Mesoscale Modeling 189
Macroscopic Scale Modeling 189
Integration of Multi-Scale Models 189
Conclusion 190
Python Code Snippet 190

30 Nanostructured LEDs and Lasers · · · · · · · 193
Introduction 193
Electro-optical Modeling of Nanostructured LEDs . 194
Laser Modeling in Nanostructured Devices 194
Simulation Techniques 195
Conclusion 196
Python Code Snippet 196

31 Functionalization of Nanostructures — 199
Introduction 199
Surface Functionalization Techniques 199
Modeling Surface Functionalization 200
1   Surface Energy Model 200
2   Quantum Mechanical Modeling 200
3   Kinetic Monte Carlo Simulations 201
Applications of Surface Functionalization 201
Conclusion 201
Python Code Snippet 202

32 Nanostructured Sensor Devices — 205
Introduction 205
Modeling Sensitivity in Nanostructured Sensors . . . 205
1   Sensing Mechanisms 206
2   Surface Functionalization 206
3   Mathematical Models 206
Modeling Selectivity in Nanostructured Sensors . . . 206
1   Multivariate Analysis 207
2   Pattern Recognition Algorithms 207
3   Data Fusion 207
Conclusion 207
Python Code Snippet 208

Chapter 1

Introduction to Nanostructured Semiconductors

In this chapter, we provide an overview of the unique properties and challenges in modeling nanostructured materials. Nanostructured semiconductors refer to materials with feature sizes ranging from 1 to 100 nanometers, where quantum confinement effects become significant. These materials exhibit novel electronic, optical, and thermal properties that can be tailored by controlling their size, shape, and composition.

Nanostructured Materials

Nanostructured materials encompass a wide range of systems such as quantum wells, quantum wires, quantum dots, and nanostructured heterojunctions. These systems are characterized by their reduced dimensions, leading to quantum confinement effects. The behavior of these nanostructures can be described using mathematical models based on the principles of quantum mechanics and statistical mechanics.

1 Quantum Wells

Quantum wells are thin layers of material sandwiched between layers of a different material, forming a heterostructure. The confinement of charge carriers in the quantum well direction leads to discrete energy levels, known as quantum well states. The electronic and optical properties of quantum wells can be accurately modeled using the effective mass approximation and solving the Schrödinger equation for the confined carriers.

2 Quantum Wires

Quantum wires are one-dimensional nanostructures with a high aspect ratio. The electron motion is highly confined within the wire, resulting in quantum effects such as quantized energy levels and wavefunction quantization. Understanding the transport and confinement effects in quantum wires is crucial for developing nanoscale electronic devices.

3 Quantum Dots

Quantum dots are zero-dimensional structures with confined carriers in all three dimensions. Due to their discrete energy levels, quantum dots exhibit size-dependent electronic and optical properties. They can be modeled using approaches such as the effective mass approximation or tight-binding models.

4 Nanostructured Heterojunctions

Nanostructured heterojunctions are interfaces between different nanomaterials, which can exhibit unique electronic and optical properties. Modeling the electron and hole dynamics across these nanoscale interfaces is essential for optimizing device performance and designing efficient energy conversion systems.

Challenges in Modeling Nanostructured Semiconductors

Modeling nanostructured semiconductors poses several challenges due to their complex nature and the need for accurate quantum mechanical descriptions. The following challenges need to be addressed when modeling nanostructured materials:

1 Size and Shape Effects

The size and shape of nanostructures strongly influence their properties. As the size decreases, quantum confinement effects become more pronounced, leading to discrete energy levels and increased surface-to-volume ratio. Modeling these size-dependent effects accurately is crucial for understanding and predicting the behavior of nanostructured materials.

2 Heterostructure Interface Quality

The quality of interfaces between different nanostructured materials significantly affects device performance. The presence of defects, strain, and interfacial states can introduce additional charge trapping and scattering mechanisms. Modeling these interface effects is essential for optimizing device efficiency.

3 Computational Complexity

Modeling nanostructured materials often requires solving complex mathematical equations, typically represented by partial differential equations such as the Schrödinger equation or the Poisson equation. These equations can be computationally demanding, especially for large-scale systems. Developing efficient numerical algorithms and computational techniques is necessary to tackle this challenge.

4 Multi-Physics Coupling

Nanostructured materials often involve the coupling of multiple physical phenomena, such as electrical, thermal, and mechanical effects. Modeling these multi-physics interactions accurately requires the development of comprehensive simulation frameworks capable of capturing the interplay between different physical processes.

5 Experimental Validation

Experimental validation plays a crucial role in validating and refining theoretical models. Comparing simulation results with experimental measurements helps to improve the accuracy and reliability of the models. Developing experimental techniques that can provide detailed information about the properties and behavior of nanostructured materials is essential in advancing the field.

In the subsequent chapters, we delve deeper into specific aspects of modeling nanostructured semiconductors, such as quantum transport, band structure engineering, photonic interactions, thermoelectric properties, and many more.

Python Code Snippet

Below is a Python code snippet that models quantum well states, computes energy levels in quantum dots, and simulates electron dynamics in nanostructured heterojunctions.

```python
import numpy as np
import matplotlib.pyplot as plt

def quantum_well_energy_levels(width, effective_mass, n_levels):
    '''
    Calculate the energy levels of a quantum well.
    :param width: Width of the quantum well in nanometers.
    :param effective_mass: Effective mass of the electron in kg.
    :param n_levels: Number of energy levels to compute.
    :return: List of calculated energy levels in eV.
    '''
    h_bar = 1.0545718e-34  # Reduced Planck's constant in J.s
    e = 1.602176634e-19    # Elementary charge in Coulombs
    energy_levels = []

    for n in range(1, n_levels + 1):
        energy = (n**2 * np.pi**2 * h_bar**2) / (2 * effective_mass
            * (width * 1e-9)**2)  # in Joules
        energy_levels.append(energy / e)  # Convert to eV

    return energy_levels

def quantum_dot_properties(radius, effective_mass):
    '''
    Calculate the bandgap and confinement energy for a quantum dot.
    :param radius: Radius of the quantum dot in nanometers.
    :param effective_mass: Effective mass of the electron in kg.
    :return: Bandgap and confinement energy in eV.
    '''
    h_bar = 1.0545718e-34  # Reduced Planck's constant in J.s
    e = 1.602176634e-19    # Elementary charge in Coulombs
    confined_energy = (h_bar**2 * np.pi**2) / (2 * effective_mass *
        (radius * 1e-9)**2)  # in Joules
    bandgap = 1.5 - 0.03 * (radius - 4)  # Example bandgap
        correction based on radius (adjust as needed)
```

```python
    return bandgap, confined_energy / e  # Convert confined energy
        to eV

def electron_dynamics_in_heterojunction(electrons_initial, time,
        decay_constant):
    '''
    Simulate electron dynamics in a heterojunction.
    :param electrons_initial: Initial number of electrons.
    :param time: Time in seconds for simulation.
    :param decay_constant: Decay constant in 1/s.
    :return: Number of electrons remaining after the given time.
    '''
    remaining_electrons = electrons_initial * np.exp(-decay_constant
        * time)
    return remaining_electrons

# Inputs for the calculations
width_quantum_well = 10  # Width of quantum well in nanometers
effective_mass_electron = 9.11e-31  # Effective mass of electron in
    kg
n_levels = 5  # Number of energy levels to compute

# Calculate quantum well energy levels
energy_levels = quantum_well_energy_levels(width_quantum_well,
    effective_mass_electron, n_levels)

# Inputs for quantum dot properties
radius_quantum_dot = 5  # Radius of quantum dot in nanometers

# Calculate bandgap and confinement energy
bandgap, confined_energy =
    quantum_dot_properties(radius_quantum_dot,
    effective_mass_electron)

# Inputs for electron dynamics simulation
initial_electrons = 1e12  # Initial number of electrons
simulation_time = 1e-6  # Time in seconds
decay_constant = 1e6  # Decay constant in 1/s

# Simulate electron dynamics
remaining_electrons =
    electron_dynamics_in_heterojunction(initial_electrons,
    simulation_time, decay_constant)

# Output results
print("Quantum Well Energy Levels (eV):", energy_levels)
print("Quantum Dot Bandgap (eV):", bandgap)
print("Quantum Dot Confinement Energy (eV):", confined_energy)
print("Remaining Electrons in Heterojunction:", remaining_electrons)
```

This code defines three functions:

- `quantum_well_energy_levels` computes the energy levels of a quantum well based on its dimensions and the effective mass of electrons.
- `quantum_dot_properties` calculates the bandgap and confinement energy for a quantum dot given its radius and the effective mass of electrons.
- `electron_dynamics_in_heterojunction` simulates the decay of electrons in a heterojunction over a specified time period.

The provided example calculates energy levels for a quantum well, determines the bandgap and confinement energy for a quantum dot, and simulates electron decay in a heterojunction, then prints the results.

Chapter 2

Quantum Wells

In this chapter, we will delve into the detailed modeling of electronic and optical properties in quantum well structures. Quantum wells are thin layers of material confined between layers of a different material, forming a heterostructure. These structures exhibit unique properties due to the quantum confinement effects, which arise from the reduced dimensions of the well.

Introduction to Quantum Wells

Quantum wells are semiconductor structures that consist of a thin layer with a lower bandgap sandwiched between two layers with higher bandgaps. The confinement of charge carriers in the quantum well direction gives rise to discrete energy levels, known as quantum well states. These confined states are crucial in determining the electronic and optical properties of quantum wells.

Let us consider a simplified one-dimensional quantum well structure, where the electron motion is confined in the z direction. The potential energy within the quantum well can be approximated by a rectangular potential well with infinite potential barriers. The time-independent Schrödinger equation describing the electron wavefunction $\psi(z)$ within the well can be written as:

$$\hat{H}\psi(z) = E\psi(z)$$

where $\hat{H}$ is the Hamiltonian operator, E is the total energy of the electron, and $\psi(z)$ is the wavefunction.

Modeling Quantum Well States

In order to determine the energy levels and wavefunctions of quantum well states, it is necessary to solve the Schrödinger equation within the well region. The effective mass approximation is commonly used to simplify the calculations. This approximation considers that the electron behaves as if it has an effective mass m^*. The Hamiltonian for the one-dimensional quantum well can then be written as:

$$\hat{H} = -\frac{\hbar^2}{2m^*}\frac{d^2}{dz^2} + V(z)$$

where $V(z)$ is the potential energy within the well. For the rectangular potential well, $V(z) = 0$ within the well region and $V(z) = \infty$ outside the well region.

By substituting the Hamiltonian into the time-independent Schrödinger equation, we obtain:

$$-\frac{\hbar^2}{2m^*}\frac{d^2}{dz^2}\psi(z) = E\psi(z)$$

Simplifying the equation further, we have:

$$\frac{d^2}{dz^2}\psi(z) = -\frac{2m^*E}{\hbar^2}\psi(z)$$

Solving the Schrödinger Equation

The solution to the Schrödinger equation within the well region is determined by imposing boundary conditions. At the boundaries of the well, the wavefunction must be continuous and its derivative must be discontinuous. This leads to the quantization of energy levels within the well.

The wavefunction within the well region can be expressed as a linear combination of plane waves:

$$\psi(z) = \sum_{n=1}^{\infty} A_n \sin(k_n z) + B_n \cos(k_n z)$$

where A_n and B_n are constants determined by the boundary conditions, and $k_n = \frac{\sqrt{2m^*E_n}}{\hbar}$ is the wavevector.

By applying the boundary conditions, the energy levels E_n and the coefficients A_n, B_n can be determined. The energy levels correspond to discrete values that depend on the well width and the effective mass of the electron.

Electronic and Optical Properties

The energy levels obtained from the solution of the Schrödinger equation allow us to understand the electronic and optical properties of quantum wells. The energy separation between the levels determines the electronic transitions that can occur in the system.

By applying an external bias or changing the material composition, the energy levels in a quantum well can be tuned. This property is utilized in various applications, such as quantum well lasers, quantum well infrared photodetectors, and quantum well field-effect transistors.

Conclusion

In this chapter, we have explored the detailed modeling of electronic and optical properties in quantum well structures. By solving the Schrödinger equation within the quantum wells, the energy levels and wavefunctions can be determined. These properties play a crucial role in understanding and designing devices based on quantum wells.

In the next chapter, we will shift our focus to another type of nanostructured semiconductor, quantum wires. Quantum wires exhibit one-dimensional confinement effects, and their transport and confinement properties will be examined in detail.Here is the section containing the Python code snippet that corresponds to the important equations and algorithms discussed in the chapter on Quantum Wells, properly wrapped in LaTeX using the minted package.

Python Code Snippet

Below is a Python code snippet that calculates the energy levels and wavefunctions of electrons in a one-dimensional quantum well, based on the modeling discussed in this chapter.

```python
import numpy as np
import matplotlib.pyplot as plt

def quantum_well_states(well_width, effective_mass, num_states=5):
    '''
    Calculate energy levels and wavefunctions in a 1D quantum well.
    :param well_width: Width of the quantum well in meters.
    :param effective_mass: Effective mass of the electron in kg.
    :param num_states: Number of energy states to calculate.
    :return: Energy levels and corresponding wavefunctions.
    '''
    hbar = 1.0545718e-34  # Reduced Planck's constant in J.s
    m_star = effective_mass  # Effective mass of the electron in kg
    energies = []
    wavefunctions = []

    # Calculate energy levels
    for n in range(1, num_states + 1):
        En = (n**2 * np.pi**2 * hbar**2) / (2 * m_star *
            well_width**2)
        energies.append(En)

        # Calculate the wavefunction: psi(z) = sqrt(2/well_width) *
            sin(n*pi*z/well_width)
        def wavefunction(z):
            return np.sqrt(2/well_width) * np.sin(n * np.pi * z /
                well_width)

        wavefunctions.append(wavefunction)

    return energies, wavefunctions

def plot_wavefunctions(well_width, wavefunctions, num_states):
    '''
    Plot the wavefunctions for the quantum well.
    :param well_width: Width of the quantum well in meters.
    :param wavefunctions: List of wavefunction functions.
    :param num_states: Number of states.
    '''
    z = np.linspace(0, well_width, 1000)  # Spatial domain

    plt.figure(figsize=(10, 6))
    for n in range(num_states):
        plt.plot(z, wavefunctions[n](z), label=f'_{n+1}(z)')

    plt.title("Wavefunctions in a 1D Quantum Well")
    plt.xlabel("Position (m)")
    plt.ylabel("Wavefunction ()")
    plt.xlim(0, well_width)
    plt.ylim(-0.5, 1.5)
    plt.axhline(0, color='black', lw=0.5, linestyle='--')
    plt.legend()
```

```python
    plt.grid()
    plt.show()

# Inputs for the calculations
well_width = 1e-9       # Width of the well in meters (1 nm)
effective_mass = 9.11e-31   # Effective mass of electron in kg

# Calculate energy levels and wavefunctions
energy_levels, wavefuncs = quantum_well_states(well_width,
↪    effective_mass)

# Output energy levels
for n, E in enumerate(energy_levels, start=1):
    print(f"Energy Level E_{n}: {E:.3e} J")

# Plotting wavefunctions
plot_wavefunctions(well_width, wavefuncs,
↪    num_states=len(energy_levels))
```

This code defines two functions:

- `quantum_well_states` calculates the energy levels and wavefunctions of electrons in a one-dimensional quantum well based on its width and the effective mass of the electron.
- `plot_wavefunctions` visualizes the computed wavefunctions for the first few quantum states within the defined well.

The example provided calculates the energy levels for the first few states and plots their corresponding wavefunctions, demonstrating the effects of quantum confinement in a quantum well.

Chapter 3

Quantum Wires

In this chapter, we will delve deep into the fascinating world of quantum wires, which are one-dimensional systems that exhibit unique transport and confinement effects. As a mathematics PhD, I will provide expert insight and explore the mathematical models used to understand and characterize these remarkable structures.

Introduction to Quantum Wires

Quantum wires are ultra-thin structures that confine electrons to move only along a single dimension. By reducing the degrees of freedom, these one-dimensional systems exhibit striking quantum mechanical phenomena that differ significantly from their bulk counterparts. Understanding the transport and confinement effects in quantum wires requires a rigorous mathematical framework.

1 One-Dimensional Schrödinger Equation

The behavior of electrons in quantum wires is described by the one-dimensional time-independent Schrödinger equation, given as:

$$\hat{H}\psi(x) = E\psi(x)$$

where $\hat{H}$ is the Hamiltonian operator, $\psi(x)$ is the electron wavefunction along the wire, and E is the corresponding energy eigenvalue.

2 Quantum Wire Potential

The potential energy within a quantum wire plays a crucial role in determining the electron behavior and transport properties. Commonly used potentials include the infinite well and harmonic potential. For simplicity, let us consider the infinite well potential, which is represented by:

$$V(x) = \begin{cases} 0, & \text{for } 0 < x < L, \\ \infty, & \text{otherwise,} \end{cases}$$

where L denotes the length of the wire.

Solving the Schrödinger Equation

To understand the transport and confinement effects in quantum wires, we must solve the one-dimensional Schrödinger equation for the given potential. For an infinite well potential, the equation takes the form:

$$-\frac{\hbar^2}{2m^*} \frac{d^2\psi}{dx^2} = E\psi$$

where $\hbar$ is the reduced Planck's constant and m^* is the effective mass of the electron.

1 Boundary Conditions

To find the wavefunction $\psi(x)$ and energy eigenvalues E, suitable boundary conditions must be imposed. For the infinite well potential, the wavefunction must be continuous within the wire, and its derivative must be discontinuous at the boundaries $x = 0$ and $x = L$.

2 Solutions for Quantum Wire States

Solving the Schrödinger equation with the given boundary conditions, we obtain a set of quantized energy levels and corresponding wavefunctions for the quantum wire system. For the infinite well potential, these solutions can be expressed as:

$$\psi_n(x) = \sqrt{\frac{2}{L}} \sin\left(\frac{n\pi x}{L}\right)$$

where n is the quantum number corresponding to the energy level and determines the number of nodes in the wavefunction.

Transport and Confinement Effects

The quantized energy levels obtained from solving the one-dimensional Schrödinger equation in quantum wires give rise to intriguing transport and confinement effects.

1 Quantum Confinement

Quantum confinement refers to the restriction of electron motion in the transverse dimensions perpendicular to the wire axis. As the width of the wire decreases, the energy levels become more tightly spaced, resulting in discrete energy states. This confinement leads to the quantization of electron motion along the wire, giving rise to an energy band structure unique to one-dimensional systems.

2 Wavefunction Localization

The wavefunctions of electrons in quantum wires are spatially confined within the wire region. As the quantum number n increases, the oscillation frequency of the wavefunction increases, resulting in increased localization within the wire. This localization effect is crucial in understanding the transport behavior and carrier properties in quantum wire devices.

3 Quantum Wire Conductance

The quantized energy levels in quantum wires have a direct impact on electron transport properties. Due to the discrete energy states, only certain energies are available for electrons to occupy, leading to a step-like behavior in the conductance. This quantization of conductance reveals the unique transport properties of quantum wires.

Conclusion

In this chapter, we have explored the mathematical foundation underlying the understanding of transport and confinement effects in one-dimensional quantum wires. By solving the one-dimensional

Schrödinger equation, we obtained the quantized energy levels and corresponding wavefunctions for quantum wire systems. These quantized levels, limited electron motion, and confinement effects play a crucial role in determining the transport behavior and electronic properties of quantum wires.

In the subsequent chapters, we will continue our exploration of nanostructured semiconductors, investigating various other fascinating aspects and properties associated with different nanostructures.

Python Code Snippet

Below is a Python code snippet that computes the wavefunctions and energy eigenvalues for quantum wires based on the infinite well potential.

```python
import numpy as np
import matplotlib.pyplot as plt

def calculate_wavefunction(n, L, x):
    '''
    Calculate the wavefunction for a quantum wire using the infinite
      well potential.
    :param n: Quantum number (1, 2, 3,...).
    :param L: Length of the quantum wire.
    :param x: Position along the wire (numpy array).
    :return: Wavefunction at positions x.
    '''
    return np.sqrt(2 / L) * np.sin(n * np.pi * x / L)

def calculate_energy(n, hbar, m_star, L):
    '''
    Calculate the energy eigenvalue for a quantum wire.
    :param n: Quantum number (1, 2, 3,...).
    :param hbar: Reduced Planck's constant.
    :param m_star: Effective mass of the electron.
    :param L: Length of the quantum wire.
    :return: Energy eigenvalue.
    '''
    return (n**2 * (np.pi**2) * hbar**2) / (2 * m_star * L**2)

# Constants
hbar = 1.0545718e-34    # Reduced Planck's constant (J·s)
m_star = 9.11e-31       # Effective mass of the electron (kg)
L = 1e-9                # Length of the quantum wire (1 nm)

# Quantum number range
n_values = np.arange(1, 6)    # First 5 quantum states
```

```python
x = np.linspace(0, L, 100)   # Position along the wire

# Plotting wavefunctions
plt.figure(figsize=(10, 6))
for n in n_values:
    psi = calculate_wavefunction(n, L, x)
    plt.plot(x * 1e9, psi, label=f'Wavefunction (n={n})')

plt.title('Wavefunctions for Quantum Wires')
plt.xlabel('Position (nm)')
plt.ylabel('Wavefunction ()')
plt.legend()
plt.grid()
plt.ylim(-1.5, 1.5)
plt.axhline(0, color='black', linewidth=0.5, ls='--')
plt.show()

# Calculate and print energy levels
energy_levels = [calculate_energy(n, hbar, m_star, L) for n in n_values]
for n, energy in zip(n_values, energy_levels):
    print(f"Energy eigenvalue for n={n}: {energy:.2e} J")
```

This code defines two functions:

- `calculate_wavefunction` calculates the wavefunction for a quantum wire based on the infinite well potential and a specified quantum number.
- `calculate_energy` computes the energy eigenvalues for given quantum states.

The provided example calculates the wavefunctions for the first five quantum states and plots them, displaying how the wavefunctions vary with position along the wire. Additionally, it computes and prints the energy eigenvalues corresponding to each quantum state.

Chapter 4

Quantum Dots

Quantum dots are zero-dimensional structures that possess unique electronic and excitonic properties due to their discrete energy levels and quantum confinement effects. In this chapter, we will explore advanced mathematical models used to study these properties in detail. As a mathematics PhD, I will provide expert insight into the development and application of these models.

Introduction to Quantum Dots

Quantum dots are nanoscale semiconductor particles with sizes on the order of a few nanometers. Unlike bulk materials, quantum dots exhibit discretized energy levels and strong quantum confinement due to their small dimensions. These characteristics give rise to unique electronic and optical properties that have led to their widespread use in various applications, such as optoelectronic devices and quantum computing.

1 Size and Shape Effects

The size and shape of quantum dots play a crucial role in determining their electronic and excitonic properties. As the size decreases, the energy levels become more closely spaced, and the influence of quantum confinement becomes more pronounced. Additionally, the shape of the quantum dot can affect its symmetry and energy spectrum, leading to variations in its properties.

2 Excitonic Properties

Excitons, which are bound states of an electron and a hole, are fundamental to understanding the optical properties of quantum dots. Due to the strong electron-hole Coulomb interaction and quantum confinement effects, excitons in quantum dots exhibit unique energy levels and optical transitions. Modeling these excitonic properties requires a comprehensive mathematical treatment.

Modeling Quantum Dots

Modeling quantum dots requires solving the Schrödinger equation within the framework of appropriate boundary conditions. Various mathematical methods have been developed to accurately describe the electronic and excitonic properties of quantum dots. In this section, we will explore some advanced models used in the field.

1 Effective Mass Approximation

The effective mass approximation is a widely used model for describing the electronic structure of quantum dots. In this approximation, the behavior of electrons is approximated by considering an effective mass that captures the average behavior of the electron within the quantum dot. The Schrödinger equation within this approximation takes the form:

$$\left(-\frac{\hbar^2}{2m^*}\nabla^2 + V(\mathbf{r})\right)\psi(\mathbf{r}) = E\psi(\mathbf{r})$$

where $\hbar$ is the reduced Planck's constant, m^* is the effective mass, $V(\mathbf{r})$ is the potential within the quantum dot, E is the energy eigenvalue, and $\psi(\mathbf{r})$ is the wavefunction.

2 Poisson-Schrödinger Equation

To accurately model the potential within a quantum dot, the Poisson equation is coupled with the Schrödinger equation, giving rise to the Poisson-Schrödinger equation:

$$-\nabla \cdot (\epsilon(\mathbf{r})\nabla\phi(\mathbf{r})) = \rho(\mathbf{r})$$

where $\epsilon(\mathbf{r})$ is the dielectric function, $\phi(\mathbf{r})$ is the electrostatic potential, and $\rho(\mathbf{r})$ is the charge density. Solving this equation self-consistently with the Schrödinger equation allows for an accurate description of the electronic and excitonic properties of quantum dots, accounting for the effects of charge distributions.

3 Density Functional Theory (DFT)

Density functional theory is a powerful computational method used to study the electronic structure and properties of quantum dots. In DFT, the many-electron problem is reduced to a one-electron problem by considering the electron density as the fundamental quantity. By solving the Kohn-Sham equations self-consistently, the ground-state properties of quantum dots, such as energy levels and wavefunctions, can be obtained.

4 Configuration Interaction (CI)

Configuration interaction is an advanced many-body technique used to include electron-electron interactions beyond the mean-field level. In CI, a basis set of single-particle states is chosen, and the many-electron wavefunction is expanded as a linear combination of these basis functions. By including electron-electron correlations through the configuration interaction method, highly accurate results for the excitonic properties of quantum dots can be obtained.

Conclusion

In this chapter, we have explored advanced mathematical models used to describe the electronic and excitonic properties of quantum dots. These models, such as the effective mass approximation, Poisson-Schrödinger equation, density functional theory, and configuration interaction, provide invaluable tools for understanding and predicting the behavior of quantum dots. By employing these models, researchers can gain insight into the unique properties of quantum dots and design novel devices with enhanced functionality.

Python Code Snippet

Below is a Python code snippet that implements various equations and algorithms related to quantum dots, including the effective mass approximation, the Poisson-Schrödinger equation, and configuration interaction.

```python
import numpy as np
from scipy.integrate import solve_bvp
from scipy.linalg import eigh

def effective_mass_approximation(mass, potential, grid_points):
    '''
    Solves the effective mass Schrödinger equation for a quantum
     ↪ dot.
    :param mass: Effective mass of the electron in kg.
    :param potential: Potential energy as a function of position.
    :param grid_points: Array of spatial points where the solution
     ↪ is computed.
    :return: Energy levels and wavefunctions.
    '''
    hbar = 1.0545718e-34  # Reduced Planck's constant in J.s
    kinetic_energy = -(hbar**2 / (2 * mass)) *
     ↪ np.diff(np.eye(len(grid_points)), 2)

    # Construct the Hamiltonian matrix
    hamiltonian = kinetic_energy + np.diag(potential)

    # Solve for eigenvalues and eigenvectors
    energies, wavefunctions = eigh(hamiltonian)
    return energies, wavefunctions

def poisson_schrodinger(potential, rho, eps, num_points):
    '''
    Solves the coupled Poisson-Schrödinger equations.
    :param potential: Initial potential guess.
    :param rho: Charge density.
    :param eps: Dielectric constant.
    :param num_points: Number of grid points.
    :return: Updated potential.
    '''
    def boundary_conditions(ya, yb):
        return np.array([ya[0], yb[0]])

    left_edge = 0.0
    right_edge = 1.0
    x = np.linspace(left_edge, right_edge, num_points)

    # Poisson equation
    def poisson_eq(x, y):
        return np.vstack((y[1], -rho(x) / eps))

    solution = solve_bvp(poisson_eq, boundary_conditions, x,
     ↪ potential)
    return solution.y[0]

def configuration_interaction(eigenstates, num_electrons):
    '''
```

```
    Calculates the configuration interaction for a given basis of
↪   eigenstates.
    :param eigenstates: Basis of single-particle eigenstates.
    :param num_electrons: Number of electrons in the system.
    :return: Total energy of the system.
    '''
    # Placeholder for interaction term computation
    energy = 0
    # Sum over pairwise interactions for all configurations
    for i in range(num_electrons):
        for j in range(i + 1, num_electrons):
            energy += np.dot(eigenstates[i], eigenstates[j])    #
↪           Example interaction term
    return energy

# Example parameters
mass = 9.11e-31   # Mass of the electron in kg
num_grid_points = 100
x_grid = np.linspace(0, 10e-9, num_grid_points)   # Spatial grid from
↪   0 to 10 nm
potential = np.zeros(num_grid_points)   # Initial potential
charge_density = np.ones(num_grid_points)   # Example uniform charge
↪   density
dielectric_constant = 11.7   # Example for GaAs

# Quantum Dots model calculations
energy_levels, wavefuncs = effective_mass_approximation(mass,
↪   potential, x_grid)
updated_potential = poisson_schrodinger(potential, lambda x:
↪   charge_density, dielectric_constant, num_grid_points)
print("Energy Levels (J):", energy_levels)
print("Updated Potential (V):", updated_potential)

# Configuration Interaction example
num_electrons = 2
total_energy = configuration_interaction(wavefuncs, num_electrons)
print("Total Energy with CI (J):", total_energy)
```

This code defines three functions:

- `effective_mass_approximation` computes the energy levels and wavefunctions using the effective mass approximation for a quantum dot.
- `poisson_schrodinger` solves the coupled Poisson-Schrödinger equations to update the potential based on charge density and the dielectric constant.
- `configuration_interaction` estimates the total energy of the system through configuration interaction using the basis of single-particle eigenstates.

In the example provided, the code computes the energy levels, updates the potential, and estimates the total energy for a simple quantum dot model, then prints the results.

Chapter 5

Nanostructured Heterojunctions

Nanostructured heterojunctions play a crucial role in modern semiconductor devices, enabling efficient charge transport and facilitating the separation of photo-generated carriers. In this chapter, we will delve into the mathematical modeling of electron and hole dynamics across nanoscale interfaces. By accurately describing the behavior of charge carriers in these systems, we can gain valuable insight into the design and optimization of heterojunction-based devices.

Introduction to Nanostructured Heterojunctions

Nanostructured heterojunctions consist of interfaces between different semiconductor materials, where the energy band structure and electron affinity change abruptly. This discontinuity in the electronic properties results in a spatial variation in the electron and hole densities across the interface. Understanding and modeling the dynamics of these charge carriers at the interface is crucial for predicting the performance of heterojunction-based devices.

1 Band Alignment at Interfaces

The energy-level alignment at a heterojunction interface determines the behavior of charge carriers near the interface. Two important concepts in band alignment are band bending and band offset. Band bending occurs due to the redistribution of charge carriers near the interface, creating an electric field that modifies the energy bands. Band offset describes the energy difference between the conduction and valence bands across the interface.

2 Charge Transport in Heterojunctions

The transport of charge carriers across nanostructured heterojunctions is governed by various mechanisms, including thermionic emission, tunneling, and recombination. Modeling and understanding these transport processes are essential for optimizing device performance. In addition, the presence of defects or impurities in the heterojunction interface can significantly influence charge transport behavior.

Mathematical Modeling of Heterojunction Interfaces

Mathematical models for nanostructured heterojunctions aim to describe the dynamics of electrons and holes across the interface by solving appropriate transport equations. In this section, we will explore some of the important mathematical frameworks used for modeling heterojunction interfaces.

1 The Drift-Diffusion Model

The drift-diffusion model is a widely used approach to describe the transport of charge carriers in nanostructured heterojunctions. This model combines the effects of carrier drift, caused by electric fields, and carrier diffusion, driven by concentration gradients. For electrons, the drift-diffusion equation takes the form:

$$\frac{\partial n}{\partial t} = \nabla \cdot (D_n \nabla n) + \mu_n n \left(\mathbf{E} + \frac{1}{q} \nabla \phi \right) + G_r - R_{nr}$$

where n is the electron density, t is time, D_n is the electron diffusion coefficient, μ_n is the electron mobility, $\mathbf{E}$ is the electric field, ϕ is the electrostatic potential, q is the elementary charge, G_r is the generation rate of electrons, and R_{nr} is the recombination rate of electrons. Similar equations can be derived for holes.

2 Schrodinger-Poisson Solver

To accurately describe the electronic properties of nanostructured heterojunctions, the Schrodinger-Poisson solver is often employed. This solver combines the Schrodinger equation, describing the quantum mechanical behavior of electrons in the presence of potential energy, with the Poisson equation, capturing the electrostatics of the heterojunction. The coupled equations can be solved self-consistently to obtain the charge density, energy bands, and wavefunctions at the heterojunction interface.

3 Kinetic Monte Carlo Simulations

Kinetic Monte Carlo simulations provide a powerful computational tool for studying charge transport in nanostructured heterojunctions. This approach models the stochastic motion of individual charge carriers through random hopping processes. By considering the relevant energy barriers and carrier concentrations, kinetic Monte Carlo simulations can capture the real-time dynamics of charge carriers, including carrier trapping, recombination, and transport across the interface.

4 Advanced Approaches

Advanced mathematical approaches, such as non-equilibrium Green's functions and tight-binding models, are also used to model electron and hole dynamics across nanostructured heterojunctions. These methods offer a more detailed description of carrier behavior, including quantum mechanical effects and electron-phonon interactions. However, they require more computational resources and are often applied in combination with other modeling techniques.

Conclusion

In this chapter, we have explored the mathematical modeling of electron and hole dynamics across nanostructured heterojunctions. By understanding the behavior of charge carriers at these interfaces, we can optimize the performance of heterojunction-based devices such as photovoltaic cells and transistors. Through the application of the drift-diffusion model, Schrodinger-Poisson solver, kinetic Monte Carlo simulations, and advanced mathematical approaches, researchers can gain valuable insights into the transport mechanisms, recombination processes, and the overall performance of nanostructured heterojunctions.

Python Code Snippet

Below is a Python code snippet that implements important equations and algorithms discussed in the context of modeling charge transport in nanostructured heterojunctions.

```python
import numpy as np

def drift_diffusion_model(n, mu_n, D_n, E, phi, G_r, R_nr,
     time_step):
    '''
    Calculate the electron density evolution using the
     Drift-Diffusion model.
    :param n: Initial electron density (numpy array).
    :param mu_n: Electron mobility (float).
    :param D_n: Electron diffusion coefficient (float).
    :param E: Electric field (numpy array).
    :param phi: Electrostatic potential (numpy array).
    :param G_r: Generation rate (numpy array).
    :param R_nr: Recombination rate (numpy array).
    :param time_step: Time step for simulation (float).
    :return: Updated electron density (numpy array).
    '''
    drift_term = mu_n * n * (E + np.gradient(phi))   # Drift
     component
    diffusion_term = D_n * np.gradient(np.gradient(n))   # Diffusion
     component

    dn_dt = diffusion_term + drift_term + G_r - R_nr   # Rate of
     change of electron density
    n += dn_dt * time_step   # Update electron density
    return n


def schrodinger_poisson_solver(potential, mass, boundary_conditions,
     num_points):
    '''
    Solve Schrodinger and Poisson equations self-consistently.
    :param potential: Initial potential array (numpy array).
    :param mass: Effective mass of the charge carriers (float).
    :param boundary_conditions: Boundary conditions for solving the
     equations.
    :param num_points: Number of grid points for the solution (int).
    :return: Energy levels and wavefunctions (numpy arrays).
    '''
    # Simplified example, assumes constant grid size and potential
```

```python
    grid = np.linspace(boundary_conditions['left'],
     ↪ boundary_conditions['right'], num_points)
    h_bar = 1.054571e-34  # Planck's constant over 2 pi
    energies = []  # To store energy levels
    wavefunctions = []  # To store wavefunctions

    for n in range(1, num_points):
        # Solve for energy and wavefunction here (simplified)
        energy = h_bar**2 * n**2 / (2 * mass)  # Energy levels
        wavefunction = np.sin(n * np.pi * grid /
         ↪ boundary_conditions['right'])  # Example wavefunction

        energies.append(energy)
        wavefunctions.append(wavefunction)

    return np.array(energies), np.array(wavefunctions)


def kinetic_monte_carlo(n_steps, barriers, carrier_concentrations):
    '''
    Perform Kinetic Monte Carlo simulation for charge transport.
    :param n_steps: Number of steps in the simulation (int).
    :param barriers: List of energy barriers for hopping processes
     ↪ (list of floats).
    :param carrier_concentrations: Concentration of carriers (numpy
     ↪ array).
    :return: Updated carrier positions (numpy array).
    '''
    positions = np.zeros((n_steps, len(carrier_concentrations)))  #
     ↪ Store positions over time

    for step in range(n_steps):
        for index in range(len(carrier_concentrations)):
            if np.random.rand() < np.exp(-barriers[index]):  #
             ↪ Simple Metropolis criterion for hopping
                positions[step, index] += np.random.choice([-1, 1])
                 ↪ # Hop left or right

    return positions


# Parameters for simulations
initial_electron_density = np.ones(100) * 1e10  # Initial electron
 ↪ density (cm^-3)
electron_mobility = 1500  # Mobility in cm^2/(V·s)
diffusion_coefficient = 36  # Diffusion coefficient in cm^2/s
electric_field = np.ones(100) * 0.1  # Uniform electric field (V/cm)
electrostatic_potential = np.zeros(100)  # Initial potential
generation_rate = np.zeros(100) + 1e20  # Constant generation rate
 ↪ (1/cm^3s)
recombination_rate = np.zeros(100)  # No recombination (for
 ↪ simplicity)
time_step = 1e-12  # Time step in seconds
```

```python
# Drift-Diffusion Model Calculation
updated_electron_density = drift_diffusion_model(
    initial_electron_density, electron_mobility,
    ↪ diffusion_coefficient, electric_field,
    electrostatic_potential, generation_rate, recombination_rate,
    ↪ time_step
)

# Schrodinger-Poisson Solver
boundary_conditions = {'left': -1e-7, 'right': 1e-7}  # Example
↪ boundaries in meters
mass = 9.11e-31  # Effective mass of electron in kg
energy_levels, wavefunctions = schrodinger_poisson_solver(
    np.zeros(100), mass, boundary_conditions, num_points=100
)

# Kinetic Monte Carlo Simulation
n_steps = 1000
barriers = np.random.uniform(0, 1, 100)  # Random energy barriers
↪ for hopping
carrier_concentrations = np.ones(100) * 1e10  # Initial
↪ concentrations
updated_positions = kinetic_monte_carlo(n_steps, barriers,
↪ carrier_concentrations)

# Output results
print("Updated Electron Density:", updated_electron_density)
print("Energy Levels:", energy_levels)
print("Wavefunctions:", wavefunctions)
print("Updated Carrier Positions (KMC):", updated_positions)
```

This code defines several functions:

- `drift_diffusion_model` simulates the evolution of electron density using the Drift-Diffusion model.
- `schrodinger_poisson_solver` approximates solutions for the Schrodinger and Poisson equations in a given potential.
- `kinetic_monte_carlo` conducts a Kinetic Monte Carlo simulation to model stochastic charge carrier transport.

The provided example initializes parameters and simulates the updated electron density, energy levels, wavefunctions, and carrier positions, then prints the results.

Chapter 6

Nano-MOSFETs

In this chapter, we will delve into the fascinating world of nano-MOSFETs (Metal-Oxide-Semiconductor Field-Effect Transistors). These devices have become the backbone of modern electronics, enabling the development of high-performance integrated circuits. We will explore the effects of scaling and quantum mechanics in nanoscale MOSFETs, shedding light on the behavior and limitations of these devices as their dimensions approach the atomic scale.

Introduction to Nano-MOSFETs

Nano-MOSFETs are key building blocks in the fabrication of nanoscale integrated circuits. They consist of a metal gate electrode, an oxide insulating layer, and a semiconducting channel. By modulating the voltage at the gate, the electric field in the channel can be controlled, allowing the conduction of current between the source and drain terminals. Nano-MOSFETs offer significant advantages in terms of size, power consumption, speed, and integration density compared to their macroscopic counterparts.

1 MOSFET Scaling

One of the most critical factors driving the development of nano-MOSFETs is scaling, which refers to the continuous reduction of device dimensions. As the size of MOSFETs decreases, device performance improves in terms of speed, power consumption, and integration density. However, scaling also introduces several fundamental challenges, such as short-channel effects, leakage currents, and device variability. Understanding and mitigating these challenges are of paramount importance in nano-MOSFET design and optimization.

2 Quantum Mechanical Effects

In nanoscale MOSFETs, quantum mechanical effects become increasingly prominent. When device dimensions approach the nanometer scale, the discreteness of charge carriers and the confinement of their wavefunctions within the channel region must be taken into account. Quantum mechanical phenomena, such as tunneling, quantum confinement, and quantum capacitance, significantly impact the behavior of the device. Therefore, accurate modeling and simulation techniques that incorporate quantum mechanics are essential for understanding and predicting the performance of nano-MOSFETs.

Mathematical Modeling of Nano-MOSFETs

Mathematical models play a crucial role in understanding the behavior of nano-MOSFETs and optimizing their performance. In this section, we will explore some of the fundamental mathematical frameworks used for modeling nano-MOSFETs.

1 Drift-Diffusion Model

The drift-diffusion model is a widely employed approach for simulating charge transport in nano-MOSFETs. It combines the effects of carrier drift, caused by electric fields, and carrier diffusion, driven by concentration gradients. The drift-diffusion equations describe the motion of charge carriers (electrons and holes) in the channel region:

$$\frac{\partial n}{\partial t} = \nabla \cdot (D_n \nabla n) - \frac{n}{\tau_n} + G_r$$

$$\frac{\partial p}{\partial t} = \nabla \cdot (D_p \nabla p) - \frac{p}{\tau_p} + G_r$$

where n and p are the electron and hole densities, D_n and D_p are the diffusion coefficients, τ_n and τ_p are the carrier lifetimes, and G_r represents carrier generation processes. These equations, coupled with Poisson's equation for the electrostatic potential, provide insights into carrier transport and device characteristics in nano-MOSFETs.

2 The Effective Mass Approximation

The Effective Mass Approximation (EMA) is often employed to simplify the quantum mechanical calculations in nano-MOSFETs. This approximation assumes that the dispersion of energy bands in the channel region is parabolic, with an effective mass parameter (m^*) characterizing the carrier dynamics. The EMA is especially useful when studying carrier transport, mobility, and density of states in nanostructured MOSFETs, as it enables the use of traditional drift-diffusion models with modified carrier masses.

3 Non-Equilibrium Green's Functions

For a more accurate description of carrier transport in nano-MOSFETs at the atomic scale, Non-Equilibrium Green's Functions (NEGF) formalism is employed. NEGF allows for the calculation of current-voltage characteristics, transmission probabilities, and the density of states by accounting for quantum mechanical effects, including wave interference and electron-electron interactions. This approach provides deeper insights into the quantum transport phenomena in nano-MOSFETs but necessitates complex mathematical formalism and computational resources.

4 Monte Carlo Simulations

Monte Carlo simulations are widely used to explore statistical and quantum mechanical effects in nano-MOSFETs. By numerically sampling carrier trajectories and considering scattering processes, Monte Carlo simulations can provide insights into carrier transport and device performance at the nanoscale. The inclusion of quantum mechanical effects, such as band structure and scattering mechanisms, allows for a more accurate description of carrier behavior, including velocity saturation, ballistic transport, and tunneling.

Conclusion

In this chapter, we have delved into the realm of nano-MOSFETs, exploring the effects of scaling and quantum mechanics on their behavior. By understanding the challenges and opportunities presented by scaling and incorporating quantum mechanical effects into mathematical models, researchers can gain deeper insights into the performance and limitations of these nanoscale devices. The drift-diffusion model, along with the effective mass approximation, NEGF formalism, and Monte Carlo simulations, provide valuable tools for exploring the intricate interplay between scaling, quantum mechanics, and device performance in nano-MOSFETs. Here is a comprehensive Python code snippet that implements important equations and algorithms mentioned in the chapter on nano-MOSFETs.

Python Code Snippet

Below is a Python code snippet that implements the drift-diffusion model, effective mass approximation, Non-Equilibrium Green's Functions (NEGF) formalism, and Monte Carlo simulations relevant to nano-MOSFETs.

```python
import numpy as np
import matplotlib.pyplot as plt
from scipy.integrate import solve_ivp

# Drift-Diffusion Model
def drift_diffusion_model(n0, p0, Dn, Dp, tau_n, tau_p, G_r,
    t_span):
    '''
    Solve the drift-diffusion equations for electron and hole
        densities.

    :param n0: Initial electron density.
    :param p0: Initial hole density.
    :param Dn: Electron diffusion coefficient.
    :param Dp: Hole diffusion coefficient.
    :param tau_n: Electron lifetime.
    :param tau_p: Hole lifetime.
    :param G_r: Generation rate.
    :param t_span: Time span for the simulation.

    :return: Electron and hole density solutions over time.
    '''
    def equations(t, y):
        n = y[0]
        p = y[1]
        dn_dt = (Dn * (np.gradient(n) + G_r)) - (n / tau_n)
        dp_dt = (Dp * (np.gradient(p) + G_r)) - (p / tau_p)
        return [dn_dt, dp_dt]
```

```python
    y_init = [n0, p0]
    sol = solve_ivp(equations, t_span, y_init, method='RK45')

    return sol.t, sol.y

# Effective Mass Approximation
def effective_mass_transport(electric_field, mass_eff):
    '''
    Calculate drift velocity using effective mass approximation.

    :param electric_field: Applied electric field in V/m.
    :param mass_eff: Effective mass of the charge carriers.

    :return: Drift velocity.
    '''
    charge = 1.6e-19  # Charge of an electron in coulombs
    mobility = charge / (mass_eff * electric_field)
    return mobility * electric_field

# Simple NEGF Implementation (Requires additional libraries for full
↪    implementation)
def negf_current(vgs, vds):
    '''
    Estimate current using NEGF formalism.

    :param vgs: Gate-to-source voltage.
    :param vds: Drain-to-source voltage.

    :return: Estimated current.
    '''
    # Placeholder for NEGF calculations (requires complex
    ↪    implementation)
    # Assumed parameters for a simple model
    kT = 0.0259  # Thermal voltage at room temperature
    g0 = 1e-6  # Conductance
    return g0 * (vgs - vds) / kT

# Monte Carlo Simulation for Carrier Transport
def monte_carlo_simulation(num_carriers, steps, dt):
    '''
    Monte Carlo simulation for carrier transport with random walk.

    :param num_carriers: Number of carriers to simulate.
    :param steps: Number of time steps.
    :param dt: Time increment for each step.

    :return: Trajectories of carriers in the channel.
    '''
    trajectories = np.zeros((num_carriers, steps))
    for i in range(num_carriers):
        position = 0
        for j in range(steps):
```

```python
            random_step = np.random.normal(0, 1)  # Random step
            position += random_step * dt  # Update position
            trajectories[i, j] = position
    return trajectories

# Example inputs for the calculations
n0 = 1e10  # Initial electron density (cm^-3)
p0 = 1e10  # Initial hole density (cm^-3)
Dn = 25  # Electron diffusion coefficient (cm^2/s)
Dp = 25  # Hole diffusion coefficient (cm^2/s)
tau_n = 1e-9  # Electron lifetime (s)
tau_p = 1e-9  # Hole lifetime (s)
G_r = 1e10  # Generation rate (cm^-3s^-1)
t_span = (0, 1e-6)  # Time span
mass_eff = 0.26 * 9.11e-31  # Effective mass of carriers (kg)
electric_field = 1e4  # Electric field (V/m)
num_carriers = 100  # Number of carriers for Monte Carlo
steps = 1000  # Number of steps for Monte Carlo
dt = 1e-12  # Time increment for Monte Carlo (s)

# Drift-Diffusion Model Calculation
time, densities = drift_diffusion_model(n0, p0, Dn, Dp, tau_n,
↪    tau_p, G_r, t_span)
electron_density = densities[0]
hole_density = densities[1]

# Effective Mass Calculation
drift_velocity = effective_mass_transport(electric_field, mass_eff)

# NEGF Current Calculation
vgs = 0.7  # Gate-to-source voltage (V)
vds = 0.5  # Drain-to-source voltage (V)
current_negf = negf_current(vgs, vds)

# Monte Carlo Simulation
trajectories = monte_carlo_simulation(num_carriers, steps, dt)

# Plotting results for Drift-Diffusion
plt.figure(figsize=(12, 6))
plt.plot(time, electron_density, label='Electron Density')
plt.plot(time, hole_density, label='Hole Density')
plt.title('Drift-Diffusion Model')
plt.xlabel('Time (s)')
plt.ylabel('Density (cm^-3)')
plt.legend()
plt.grid()
plt.show()

# Display results
print("Drift Velocity:", drift_velocity, "m/s")
print("Estimated Current (NEGF):", current_negf, "A")
print("Monte Carlo Trajectory for First Carrier:", trajectories[0])
```

This code defines several functions:

- `drift_diffusion_model` calculates the evolution of electron and hole densities over time using the drift-diffusion model.
- `effective_mass_transport` computes the drift velocity of charge carriers based on effective mass and electric field.
- `negf_current` provides a placeholder for calculating current using a simple NEGF model.
- `monte_carlo_simulation` simulates carrier transport using random walk in a Monte Carlo manner.

The provided example calculates the electron and hole densities in a nano-MOSFET, the drift velocity, current using NEGF, and simulates the carrier trajectories, also displaying the drift-diffusion model results graphically.

Chapter 7

Band Structure Engineering

In this chapter, we will explore the fascinating field of band structure engineering in nanostructures. Band structure engineering refers to the design and computational methods used to tailor the electronic band structure of materials at the nanoscale. By manipulating the band structure, scientists and engineers can control the electrical and optical properties of materials, opening up new possibilities for the development of novel devices with improved performance and functionality.

Introduction to Band Structure Engineering

The electronic band structure of a material is a key factor in determining its electronic and optical properties. In bulk materials, the band structure is largely determined by the crystal structure and the interactions between atoms. However, at the nanoscale, the confinement of electrons and the presence of interfaces and surfaces introduce additional degrees of freedom for engineering the band structure.

Band structure engineering aims to control the energy dispersion and bandgap of materials, enabling the customization of their electronic and optical properties. By tailoring the band structure, it is possible to enhance charge carrier mobility, manipulate the absorption and emission of light, and create materials with desired properties for specific applications, such as optoelectronics, energy conversion, and quantum computing.

1 Methods for Band Structure Engineering

Band structure engineering can be achieved through various methods, including:

- **Strain Engineering:** Applying mechanical strain to a material can modify its band structure. Strain changes the interatomic distances and alters the electron-electron interactions, leading to changes in the band energies and bandgap.

- **Doping:** Introducing impurities (dopants) into a material can create additional electronic states within the band structure, modifying its electronic properties. Doping can be used to increase or decrease the carrier concentration and modulate the bandgap.

- **Heterostructures:** Combining materials with different band structures at interfaces can create new electronic states and modify the overall band structure. Heterostructures enable the engineering of quantum wells, wires, and dots, offering enhanced control over carrier confinement and transport.

- **Quantum Confinement:** Confining electrons in nanoscale dimensions can lead to quantization effects and the formation of discrete energy levels, known as quantum confinement. By controlling the size and shape of nanostructures, such as quantum dots and nanowires, the band structure can be engineered to exhibit desired electronic properties.

- **External Fields:** Applying external electric or magnetic fields can modify the band structure through the Stark effect or the Zeeman effect, respectively. These effects shift the energy levels and alter their spacing, influencing the bandgap and other band characteristics.

Computational Methods for Band Structure Engineering

Designing nanostructures with tailored band structures requires advanced computational methods that accurately model electronic properties at the atomic scale. In this section, we will discuss some of the commonly used computational techniques for band structure engineering.

1 Tight-Binding Method

The tight-binding method is a widely adopted approach for modeling the band structure of nanostructures. It provides a good balance between accuracy and computational efficiency. The tight-binding model describes the electronic structure by considering only the local interactions between neighboring atoms. This method requires knowledge of the atomic orbitals and their overlap integrals, which can be determined from first principles calculations or empirical parameters. By incorporating the tight-binding parameters into the Hamiltonian matrix, the band structure of the nanostructure can be obtained through diagonalization.

The tight-binding method is particularly suitable for systems with strong localization of electronic states, such as carbon nanotubes and graphene nanoribbons, where the electronic properties are governed by the interactions between adjacent atoms.

2 Density Functional Theory

Density Functional Theory (DFT) is a powerful computational technique used to describe the electronic structure of materials. DFT numerically solves the many-body Schrödinger equation by considering the electron density instead of the wavefunctions. The electron density is determined by minimizing the total energy of the system with respect to the electron density, using an effective potential derived from the exchange-correlation functional.

DFT provides accurate predictions of the electronic band structure of materials, including their band energies and bandgap. It is widely employed for studying the properties of nanostructures, such as quantum dots, nanowires, and thin films. The versatility of DFT allows for the investigation of various band structure engineering techniques and their impact on the electronic properties of materials.

3 Empirical Pseudopotential Method

The empirical pseudopotential method is a widely used approximation for calculating the electronic band structure of materials. This method simplifies the interactions between electrons and atomic cores by replacing the complex many-body interactions with pseudopotentials. Empirical pseudopotentials are constructed to reproduce the electronic structure and properties of the material of interest.

The empirical pseudopotential method is computationally efficient and widely applicable to a broad range of materials. It enables the study of larger systems and longer time scales compared to more computationally demanding methods, such as the tight-binding method or DFT. However, the empirical pseudopotential method may introduce some degree of inaccuracy due to the approximations made in constructing the pseudopotentials.

4 Multi-Scale Modeling Approaches

Band structure engineering often requires a multi-scale modeling approach, combining different computational methods to capture the electronic properties at various length and time scales. Multi-scale modeling techniques involve bridging the results obtained from atomistic models, such as tight-binding or DFT, with continuum models, such as effective mass models or drift-diffusion models. This approach enables the investigation of band structure engineering effects on device performance at larger scales.

By integrating different modeling techniques, researchers can gain a comprehensive understanding of how the tailored band structure influences the electronic and optical properties of nanostructures, as well as their functional performance in devices.

Conclusion

In this chapter, we explored band structure engineering in nanostructures, focusing on the design and computational methods used to tailor the band structure. Band structure engineering offers a powerful approach to customize the electronic and optical properties of materials at the nanoscale, enabling new possibilities for device applications. By manipulating the band structure through methods such as strain engineering, doping, heterostructures, quantum confinement, and external fields, researchers can achieve desired electronic properties and enhance device functionality.

Advanced computational methods, including the tight-binding method, density functional theory, empirical pseudopotential method, and multi-scale modeling approaches, play a crucial role in understanding and predicting the effects of band structure engineering. These computational tools provide valuable insights into the electronic properties of nanostructures and guide the design of materials with tailored band structures.

The next chapter will delve into the modeling of nanostructured photonic crystals, exploring their light-matter interactions and defect modes. Photonic crystals present unique opportunities for manipulating light at the nanoscale and have diverse applications in areas such as optical communication, sensing, and integrated photonics.Here's the Python code snippet containing important equations and algorithms discussed in the chapter on band structure engineering. It includes functions for computing the bandgap modification due to strain, estimating the change in energy levels due to doping, and simulating the quantum confinement effect in nanostructures.

Python Code Snippet

Below is a Python code snippet for calculating the effects of strain, doping, and quantum confinement on the band structure of nanostructured semiconductor materials.

```
def bandgap_modification(strain, original_bandgap):
    '''
    Calculate the modified bandgap due to applied strain.
    :param strain: The applied strain (tensile or compressive).
    :param original_bandgap: The original bandgap of the material in
    ↪   eV.
    :return: The modified bandgap in eV.
    '''
    strain_coefficient = 0.1  # Assumed coefficient for bandgap
    ↪   strain sensitivity
    modified_bandgap = original_bandgap - strain_coefficient *
    ↪   strain
    return modified_bandgap


def doping_effect(carrier_concentration, original_bandgap):
    '''
    Estimate the change in bandgap due to doping concentration.
```

```python
    :param carrier_concentration: The carrier concentration in
    ↪  cm^-3.
    :param original_bandgap: The original bandgap of the material in
    ↪  eV.
    :return: The modified bandgap in eV.
    '''
    beta = 0.01  # Assumed coefficient for bandgap reduction per
    ↪  carrier
    change_in_bandgap = beta * (carrier_concentration * 1e-6)  #
    ↪  Convert from cm^-3 to m^-3
    modified_bandgap = original_bandgap - change_in_bandgap
    return modified_bandgap


def quantum_confinement_effect(nanostructure_size,
↪  original_bandgap):
    '''
    Calculate the increase in bandgap due to quantum confinement
    ↪  effects.
    :param nanostructure_size: Size of the nanostructure in
    ↪  nanometers.
    :param original_bandgap: The original bandgap of the material in
    ↪  eV.
    :return: The modified bandgap in eV due to quantum confinement.
    '''
    confinement_coefficient = 0.1  # Assumed coefficient for
    ↪  confinement effect
    modified_bandgap = original_bandgap + confinement_coefficient /
    ↪  nanostructure_size
    return modified_bandgap


# Inputs for the calculations
original_bandgap = 1.1  # Original bandgap of Silicon in eV
strain = 0.02  # Applied strain (2%)
carrier_concentration = 1e16  # Carrier concentration in cm^-3
nanostructure_size = 5  # Size of the nanostructure in nanometers

# Calculations
modified_bandgap_due_to_strain = bandgap_modification(strain,
↪  original_bandgap)
modified_bandgap_due_to_doping =
↪  doping_effect(carrier_concentration, original_bandgap)
modified_bandgap_due_to_quantum_confinement =
↪  quantum_confinement_effect(nanostructure_size, original_bandgap)

# Output results
print("Modified Bandgap due to Strain:",
↪  modified_bandgap_due_to_strain, "eV")
print("Modified Bandgap due to Doping:",
↪  modified_bandgap_due_to_doping, "eV")
```

```
print("Modified Bandgap due to Quantum Confinement:",
    modified_bandgap_due_to_quantum_confinement, "eV")
```

This code includes three functions:

- **bandgap_modification** calculates the modified bandgap based on the applied strain.
- **doping_effect** estimates the change in bandgap due to carrier concentration from doping.
- **quantum_confinement_effect** determines the increase in bandgap resulting from quantum confinement effects in nanostructures.

The provided example calculates the modified bandgap due to strain, doping, and quantum confinement, and then prints the results.

Chapter 8

Nanostructured Photonic Crystals

Introduction

In this chapter, we will delve into the fascinating world of nanostructured photonic crystals. Photonic crystals are periodic structures that exhibit a photonic bandgap, which controls the propagation of electromagnetic waves at specific frequencies or wavelengths. Nanostructuring these photonic crystals offers enhanced control over their optical properties, enabling the manipulation of light-matter interactions and the creation of defect modes within the photonic bandgap. In this section, we will provide an overview of the chapter's focus on modeling these light-matter interactions and defect modes in photonic crystal structures.

Fundamentals of Photonic Crystals

Before diving into the modeling of light-matter interactions in nanostructured photonic crystals, we must first establish the fundamentals of photonic crystals. A photonic crystal is a periodic arrangement of dielectric or metallic materials with a periodicity that is on the order of the wavelength of light. This periodicity gives rise to the formation of a photonic bandgap, a range of frequencies or wavelengths in which the propagation of light is prohibited. The bandgap arises due to the interference of multiple scattering events within the crystal lattice, leading to destructive interference at specific frequencies.

The photonic bandgap can be tailored by varying the periodicity, dielectric contrast, and structural parameters of the photonic crystal. Nanostructuring these photonic crystals further enhances the control over their bandgap properties. Additionally, the presence of defects or irregularities in the crystal lattice can lead to the formation of localized states within the bandgap, known as defect modes. These defect modes can have unique optical properties and can be exploited for various applications.

Modeling Light-Matter Interactions

To accurately describe the behavior of light in nanostructured photonic crystals, advanced computational modeling techniques are required. In this section, we will discuss the mathematical models used to simulate light-matter interactions in photonic crystal structures.

1 Maxwell's Equations

The interaction between light and matter in photonic crystals is governed by Maxwell's equations, the fundamental equations of classical electromagnetism. In their differential form, Maxwell's equations are given by:

$$\nabla \cdot \mathbf{E} = \frac{\rho}{\epsilon_0}, \quad \nabla \cdot \mathbf{B} = 0, \tag{8.1}$$

$$\nabla \times \mathbf{E} = -\frac{\partial \mathbf{B}}{\partial t}, \quad \nabla \times \mathbf{B} = \mu_0 \mathbf{J} + \mu_0 \epsilon_0 \frac{\partial \mathbf{E}}{\partial t}, \tag{8.2}$$

where $\mathbf{E}$ and $\mathbf{B}$ are the electric and magnetic fields, respectively, ρ is the charge density, ϵ_0 is the vacuum permittivity, μ_0 is the vacuum permeability, and $\mathbf{J}$ is the current density.

Solving Maxwell's equations for the electric and magnetic fields allows us to study the behavior of light within photonic crystal structures and its interaction with various materials and defects.

2 Finite Difference Time Domain (FDTD) Method

One widely used method for numerically solving Maxwell's equations is the finite difference time domain (FDTD) method. The

FDTD method discretizes space and time and approximates the derivatives in Maxwell's equations using finite differences. By iteratively updating the electric and magnetic fields over each time step, the FDTD method can simulate the propagation of light in photonic crystals and its interaction with defects.

The FDTD method is particularly effective for modeling photonic crystal structures with complex geometries and material distributions. It has been successfully employed to study various phenomena, such as waveguiding, scattering, and resonant phenomena within photonic crystals.

Defect Modes in Photonic Crystals

One of the key features of nanostructured photonic crystals is the presence of defect modes within the photonic bandgap. These defect modes arise from localized states associated with structural irregularities or intentional defects introduced into the crystal lattice. In this section, we will discuss the mathematical methods used to study defect modes in photonic crystals.

1 Plane Wave Expansion Method

The plane wave expansion (PWE) method is a powerful approach for calculating the photonic band structure and defect modes of photonic crystals. The PWE method expands the electromagnetic fields in terms of Bloch waves within the crystal lattice. By imposing appropriate boundary conditions, the eigenfrequencies and corresponding eigenmodes of the crystal can be obtained.

In the presence of a defect, additional localized modes may appear within the bandgap. These defect modes can be determined by solving an eigenvalue problem, which involves finding the frequencies and field distributions associated with the defect modes. The PWE method allows us to study the characteristics of these defect modes, such as their frequencies, quality factors, and field distributions.

Conclusion

In this chapter, we explored the modeling of light-matter interactions and defect modes in nanostructured photonic crystals. By utilizing advanced computational techniques, such as the FDTD method and the PWE method, we can accurately simulate the behavior of light in photonic crystal structures and gain insights into their optical properties.

The interaction between light and matter in photonic crystals is governed by Maxwell's equations, which describe the propagation of electromagnetic waves. Numerically solving these equations using methods like the FDTD method allows us to study the behavior of light within photonic crystals and its interaction with various materials and defects.

Nanostructured photonic crystals exhibit defect modes within the photonic bandgap, which can be calculated using the PWE method. These defect modes provide opportunities for tailoring the optical properties of photonic crystals and can be exploited for various applications, such as waveguiding, lasing, and sensing.

In the next chapter, we will shift our focus to the advanced modeling techniques used for simulating the thermal and electrical transport in nanostructured thermoelectrics.

Python Code Snippet

Below is a Python code snippet that implements the essential equations and algorithms discussed in the chapter, specifically related to modeling light-matter interactions in nanostructured photonic crystals.

```
import numpy as np
from scipy.constants import epsilon_0, mu_0, c

def solve_maxwells_equations(E0, B0, time_steps, dt):
    '''
    Simulate light propagation using Maxwell's equations in a
    ↪ discretized format.
    :param E0: Initial electric field distribution (numpy array).
    :param B0: Initial magnetic field distribution (numpy array).
    :param time_steps: Number of time steps for the simulation.
    :param dt: Time step size in seconds.
    :return: Time-evolved electric and magnetic field distributions
    ↪ (numpy arrays).
    '''
    E = E0.copy()
    B = B0.copy()
    for t in range(time_steps):
        # Update magnetic field B using the electric field E
        B = B + (dt / mu_0) * np.cross(np.gradient(E), axis=0)
        # Update electric field E using the magnetic field B
```

```python
        E = E - (dt / epsilon_0) * np.cross(np.gradient(B), axis=0)
    return E, B

def calculate_skin_depth(frequency):
    '''
    Calculate the skin depth based on the operating frequency.
    :param frequency: Frequency in hertz.
    :return: Skin depth in meters.
    '''
    wave_velocity = c  # Velocity of light in vacuum
    skin_depth = wave_velocity / (frequency * np.pi) ** 0.5
    return skin_depth

def calculate_defect_modes(photonic_crystal_structure):
    '''
    Calculate defect modes in a photonic crystal using Plane Wave
    ↪ Expansion.
    :param photonic_crystal_structure: Array representing the
    ↪ dielectric distribution of the structure.
    :return: Frequencies and mode shapes associated with defect
    ↪ modes.
    '''
    # Placeholder for implementation: Use a formal eigenvalue solver
    from scipy.linalg import eigh
    # Generating a fictitious Hamiltonian matrix based on the
    ↪ structure
    hamiltonian =
    ↪ np.random.rand(photonic_crystal_structure.shape[0],
    ↪ photonic_crystal_structure.shape[0])
    eigenvalues, eigenvectors = eigh(hamiltonian)
    return eigenvalues, eigenvectors

# Inputs for the simulations
E0 = np.zeros((100, 100))  # Initial electric field distribution
B0 = np.zeros((100, 100))  # Initial magnetic field distribution
time_steps = 1000  # Number of time steps
dt = 1e-9  # Time step size in seconds
frequency = 1e12  # Frequency in Hz (1 THz)

# Simulating light propagation
E_final, B_final = solve_maxwells_equations(E0, B0, time_steps, dt)

# Calculating skin depth
skin_depth = calculate_skin_depth(frequency)

# Dummy photonic crystal structure for defect mode calculation
photonic_crystal_structure = np.array([[1, 2, 1], [2, 1, 2], [1, 2,
↪ 1]])  # Example dielectric distribution

# Calculating defect modes
defect_frequencies, defect_modes =
↪ calculate_defect_modes(photonic_crystal_structure)
```

```
# Output results
print("Final Electric Field Distribution:\n", E_final)
print("Final Magnetic Field Distribution:\n", B_final)
print("Skin Depth:", skin_depth, "meters")
print("Defect Frequencies:\n", defect_frequencies)
print("Defect Modes Shapes:\n", defect_modes)
```

This code defines three functions:

- `solve_maxwells_equations` simulates the propagation of light by solving Maxwell's equations in a discretized format.
- `calculate_skin_depth` computes the skin depth based on the operating frequency.
- `calculate_defect_modes` calculates the defect modes in a photonic crystal using a simplified eigenvalue solver for demonstration purposes.

The provided example initializes electric and magnetic fields, computes the time evolution under Maxwell's equations, calculates the skin depth for a given frequency, and finds defect modes in a dummy photonic crystal structure before printing the results.

Chapter 9

Thermoelectric Nanostructures

Thermoelectric materials have garnered significant attention due to their ability to convert waste heat into useful electrical energy. In recent years, nanostructured thermoelectrics have emerged as promising candidates for enhanced thermoelectric performance. These materials, characterized by reduced dimensions and increased interface areas, exhibit unique and tunable thermal and electrical transport properties. In this chapter, we will explore advanced mathematical models utilized in the study of thermal and electrical transport in nanostructured thermoelectrics.

Introduction

Nanostructured thermoelectrics offer intriguing possibilities for improving energy conversion efficiency by manipulating thermal and electrical transport processes. The reduction in characteristic length scales in these materials introduces quantum confinement effects, enhanced phonon scattering, and increased power factor. Consequently, advanced modeling approaches are required to capture the complex interplay between thermal and electrical transport phenomena.

In this section, we provide an overview of the chapter's focus, which centers on advancing the understanding of thermal and electrical transport in nanostructured thermoelectrics. We will explore the mathematical models employed, including quantum transport formalisms, phonon transport equations, and the coupled transport equations.

Quantum Transport Modeling

Nanostructured thermoelectrics often exhibit electron and phonon transport behavior that deviates from their bulk counterparts due to quantum confinement effects. To describe these phenomena accurately, quantum transport models are employed.

1 Boltzmann Transport Equation

The Boltzmann transport equation is a fundamental equation used to model electron transport in materials by accounting for carrier scattering mechanisms. In the presence of external fields, the Boltzmann transport equation can be written as:

$$\frac{\partial f}{\partial t} + v \cdot \nabla_x f + F \cdot \nabla_v f = \left(\frac{df}{dt}\right)_{coll}$$

where f is the electron distribution function, v is the electron velocity, F is the external force acting on the electrons, and $\left(\frac{df}{dt}\right)_{coll}$ represents the electron collision term.

Solving the Boltzmann transport equation provides insights into electron transport phenomena, such as electron mobility, diffusion, and the Seebeck coefficient.

2 Density Functional Theory

Density functional theory (DFT) is widely employed in the study of electronic structure and transport properties in nanostructured thermoelectrics. This quantum mechanical approach enables the determination of electron energies, densities, and wave functions by solving the many-electron Schrödinger equation. Within DFT, the electron transport properties can be calculated using approaches like the non-equilibrium Green's function (NEGF) method.

The NEGF method provides a framework for calculating electron transmission and conductance through nanoscale devices. By considering the electronic structure of the nanostructure and its coupling to external electrodes, the NEGF method enables the modeling of quantum transport phenomena in thermoelectric systems.

Phonon Transport Modeling

In nanostructured thermoelectrics, modulation of lattice vibrations (phonons) plays a crucial role in achieving high thermoelectric performance. Advanced modeling techniques are employed to capture the intricate dynamics of phonon transport in these materials.

1 Boltzmann Transport Equation for Phonons

The Boltzmann transport equation can also be adapted to describe phonon transport in nanostructures. This equation, referred to as the Boltzmann transport equation for phonons, considers phonon dispersion, scattering mechanisms, and anharmonic interactions.

The phonon Boltzmann transport equation can be expressed as:

$$\frac{\partial \mathbf{g}}{\partial t} + \mathbf{v}_{\text{ph}} \cdot \nabla_x \mathbf{g} = \left(\frac{d\mathbf{g}}{dt}\right)_{\text{coll}}$$

where $\mathbf{g}$ is the phonon distribution function, $\mathbf{v}_{\text{ph}}$ is the phonon group velocity, and $\left(\frac{d\mathbf{g}}{dt}\right)_{\text{coll}}$ represents the phonon collision term.

Solving the phonon Boltzmann transport equation provides information on phonon transport properties, such as thermal conductivity, phonon scattering rates, and spectral energy distribution.

2 Molecular Dynamics Simulations

To capture complex phonon-phonon interactions and heat transport in nanostructured thermoelectrics accurately, molecular dynamics (MD) simulations are frequently utilized. MD simulations track the motion of individual atoms within the material by solving Newton's equations of motion with appropriate interatomic potentials.

By simulating the interactions between atoms and quantifying the resulting energy exchange, MD simulations provide valuable insights into heat conduction and phonon interactions in nanostructured thermoelectrics. These simulations can help predict thermal conductivity, phonon scattering mechanisms, and the lattice thermal conductivity reduction due to grain boundaries and interfaces.

Coupled Transport Modeling

To comprehensively capture the thermoelectric performance of nanostructured materials, it is crucial to consider the coupled transport of electrons and phonons. The interactions between these carriers significantly influence thermoelectric transport properties and must be accurately modeled.

1 Coupled Electron-Phonon Transport Equations

The coupled electron-phonon transport equations describe the simultaneous transport of electrons and phonons and their mutual interactions. These equations are usually solved self-consistently to obtain the steady-state or transient thermoelectric response of the material.

The coupled transport equations can be written as:

$$\frac{\partial f}{\partial t} + v \cdot \nabla_x f + F \cdot \nabla_v f = \left(\frac{df}{dt}\right)^e_{\text{coll}} + \left(\frac{df}{dt}\right)^{e\text{-}ph}_{\text{coll}},$$

$$\frac{\partial g}{\partial t} + v_{\text{ph}} \cdot \nabla_x \mathbf{g} = \left(\frac{d\mathbf{g}}{dt}\right)^{ph}_{\text{coll}} - \left(\frac{d\mathbf{g}}{dt}\right)^{ph\text{-}e}_{\text{coll}}.$$

where f is the electron distribution function, $\mathbf{g}$ is the phonon distribution function, and the collision terms $\left(\frac{df}{dt}\right)^e_{\text{coll}}$, $\left(\frac{df}{dt}\right)^{e\text{-}ph}_{\text{coll}}$, $\left(\frac{d\mathbf{g}}{dt}\right)^{ph}_{\text{coll}}$, and $\left(\frac{d\mathbf{g}}{dt}\right)^{ph\text{-}e}_{\text{coll}}$ represent electron-electron, electron-phonon, phonon-phonon, and phonon-electron collision terms, respectively.

Solving the coupled transport equations allows for an in-depth understanding of the interplay between electron and phonon transport, thereby enabling optimization of thermoelectric efficiency.

Conclusion

In this chapter, we have explored advanced mathematical models utilized in the study of thermal and electrical transport in nanostructured thermoelectrics. By employing quantum transport modeling, phonon transport modeling, and coupled transport modeling, we can gain profound insights into the intricacies of transport phenomena in these materials. These models pave the way for optimizing the design and performance of nanostructured thermoelectric materials for enhanced energy conversion efficiency.

In the next chapter, we will delve into the exciting field of plasmonic nanostructures, elucidating the simulation techniques employed for studying plasmon-enhanced phenomena and their applications.

Python Code Snippet

Below is a Python code snippet that implements key equations and algorithms from the analysis of thermal and electrical transport in nanostructured thermoelectrics.

```python
import numpy as np

def boltzmann_transport_electron(f, v, F, collision_term):
    '''
    Solve the Boltzmann transport equation for electrons.
    :param f: Electron distribution function.
    :param v: Electron velocity.
    :param F: External force acting on the electrons.
    :param collision_term: Electron collision term.
    :return: Updated electron distribution function.
    '''
    return f + (v * np.gradient(f) + F * np.gradient(f, v)) -
        collision_term

def boltzmann_transport_phonon(g, v_ph, collision_term):
    '''
    Solve the Boltzmann transport equation for phonons.
    :param g: Phonon distribution function.
    :param v_ph: Phonon group velocity.
    :param collision_term: Phonon collision term.
    :return: Updated phonon distribution function.
    '''
    return g + (v_ph * np.gradient(g)) - collision_term

def density_functional_theory(positions, potentials):
    '''
    Perform a simplified density functional theory calculation.
    :param positions: Atomic positions in the material.
    :param potentials: Interactions potentials for the system.
    :return: Electron energy levels and densities.
    '''
    # Here we will use a mock-up function, in practice, this would
        involve complex calculations.
    energies = np.array([np.sum(potentials) / pos for pos in
        positions])
    densities = np.exp(-energies)  # Simplified density calculation
    return energies, densities
```

```python
# Example parameters for the calculations
electron_dist_function = np.random.rand(100)  # Random initial
↪ electron distribution
phonon_dist_function = np.random.rand(100)  # Random initial phonon
↪ distribution
electron_velocity = 1.0  # Example electron velocity
external_force = 0.1  # Example external force
electron_collision_term = 0.01  # Example collision term for
↪ electrons
phonon_group_velocity = 1.0  # Example phonon velocity
phonon_collision_term = 0.01  # Example collision term for phonons
atomic_positions = np.random.rand(5)  # Random atomic positions
interaction_potentials = np.random.rand(5)  # Random interaction
↪ potentials

# Perform the calculations
updated_electron_dist =
↪ boltzmann_transport_electron(electron_dist_function,
↪ electron_velocity, external_force, electron_collision_term)
updated_phonon_dist =
↪ boltzmann_transport_phonon(phonon_dist_function,
↪ phonon_group_velocity, phonon_collision_term)
energy_levels, densities =
↪ density_functional_theory(atomic_positions,
↪ interaction_potentials)

# Output results
print("Updated Electron Distribution Function:",
↪ updated_electron_dist)
print("Updated Phonon Distribution Function:", updated_phonon_dist)
print("Energy Levels:", energy_levels)
print("Densities:", densities)
```

This code defines three functions:

- `boltzmann_transport_electron` implements the Boltzmann transport equation for electrons, updating the electron distribution based on velocity, an external force, and collision terms.
- `boltzmann_transport_phonon` adapts the Boltzmann transport equation for phonon distribution, similarly reflecting their dynamics through phonon group velocity and collision events.
- `density_functional_theory` performs a simplified calculation based on density functional theory to determine electron energy levels and densities based on atomic positions and interaction potentials.

The provided example initializes random distributions for electrons and phonons, calculates the updated distributions, and estimates energy levels and densities, then prints the results.

Chapter 10

Plasmonic Nanostructures

Plasmonic nanostructures have revolutionized the field of nanophotonics by enabling the manipulation of light at the nanoscale. These structures, characterized by their ability to support surface plasmon polaritons (SPPs), exhibit unique optical properties due to the strong coupling between photons and electrons. In this chapter, we delve into the simulation of plasmon-enhanced phenomena and their applications, shedding light on the underlying mathematical and computational techniques employed in this exciting field.

Introduction

The field of plasmonics focuses on the utilization of surface plasmons, which are collective oscillations of electrons at the interface between a metal and a dielectric medium. Plasmonic nanostructures, commonly fabricated from noble metals such as gold and silver, provide a versatile platform for manipulating light's propagation, localization, and enhancement at the nanoscale.

In this section, we introduce the focus of this chapter, which is simulating plasmon-enhanced phenomena and exploring their applications. We provide an overview of the mathematical and computational techniques employed in this field, including Maxwell's equations, the finite-difference time-domain (FDTD) method, and the Boundary Element Method (BEM).

Maxwell's Equations and Plasmonics

To simulate the behavior of plasmonic nanostructures, it is essential to understand the underlying electromagnetic phenomena. Maxwell's equations serve as the foundation for describing the interaction of light with matter, including the behavior of surface plasmons.

In the frequency domain, Maxwell's equations can be written as:

$$\nabla \times \mathbf{E} = -\mu \frac{\partial \mathbf{H}}{\partial t}, \tag{10.1}$$

$$\nabla \times \mathbf{H} = \epsilon \frac{\partial \mathbf{E}}{\partial t} + \mathbf{J}, \tag{10.2}$$

$$\nabla \cdot \mathbf{E} = \frac{\rho}{\epsilon}, \tag{10.3}$$

$$\nabla \cdot \mathbf{H} = 0, \tag{10.4}$$

where $\mathbf{E}$ and $\mathbf{H}$ are the electric and magnetic fields, respectively, ρ is the charge density, ϵ and μ are the permittivity and permeability of the medium, and $\mathbf{J}$ is the current density. Equations (10.1) and (10.2) represent Faraday's law and Ampere's law, respectively, while Equations (10.3) and (10.4) represent the Gauss's law for electric fields and the Gauss's law for magnetic fields.

For plasmonic nanostructures, it is essential to consider the presence of metal and dielectric regions. The permittivity and permeability in Equations (10.1) and (10.2) are given by:

$$\epsilon = \epsilon_r \epsilon_0, \tag{10.5}$$

$$\mu = \mu_0, \tag{10.6}$$

where ϵ_r is the relative permittivity of the material, ϵ_0 is the vacuum permittivity, and μ_0 is the vacuum permeability.

Finite-Difference Time-Domain (FDTD) Method

The Finite-Difference Time-Domain (FDTD) method is a widely used numerical technique for solving Maxwell's equations in both time and space domains. It discretizes the space and time coordinates and approximates the derivatives using finite differences.

In the FDTD method, Maxwell's equations are discretized as follows:

$$\frac{E_x^{n+1/2}(i,j,k) - E_x^{n-1/2}(i,j,k)}{\Delta t} = \frac{1}{\epsilon}\left(\frac{H_z^n(i,j+\frac{1}{2},k) - H_z^n(i,j-\frac{1}{2},k)}{\Delta y}\right. \tag{10.7}$$

$$\left. -\frac{H_y^n(i,j,k+\frac{1}{2}) - H_y^n(i,j,k-\frac{1}{2})}{\Delta z}\right), \tag{10.8}$$

$$\frac{H_y^{n+1/2}(i,j,k) - H_y^{n-1/2}(i,j,k)}{\Delta t} = \frac{1}{\mu}\left(\frac{E_x^n(i,j,k+\frac{1}{2}) - E_x^n(i,j,k-\frac{1}{2})}{\Delta z}\right. \tag{10.9}$$

$$\left. -\frac{E_z^n(i+\frac{1}{2},j,k) - E_z^n(i-\frac{1}{2},j,k)}{\Delta x}\right), \tag{10.10}$$

$$\frac{H_z^{n+1/2}(i,j,k) - H_z^{n-1/2}(i,j,k)}{\Delta t} = \frac{1}{\mu}\left(\frac{E_y^n(i+\frac{1}{2},j,k) - E_y^n(i-\frac{1}{2},j,k)}{\Delta x}\right. \tag{10.11}$$

$$\left. -\frac{E_x^n(i,j+\frac{1}{2},k) - E_x^n(i,j-\frac{1}{2},k)}{\Delta y}\right), \tag{10.12}$$

where $E_x^{n+1/2}(i,j,k)$, $E_y^{n+1/2}(i,j,k)$, and $E_z^{n+1/2}(i,j,k)$ represent the electric field components at time $t = (n+1/2)\Delta t$ and grid point (i,j,k). Similarly, $H_x^{n+1/2}(i,j,k)$, $H_y^{n+1/2}(i,j,k)$, and $H_z^{n+1/2}(i,j,k)$ represent the magnetic field components at the same time and grid point. Here, Δt is the time step, and Δx, Δy, and Δz are the spatial step sizes in the respective dimensions.

By iteratively solving the above equations, one can simulate the time evolution of electromagnetic fields inside plasmonic nanostructures.

Boundary Element Method (BEM)

The Boundary Element Method (BEM) is another numerical technique that is commonly used for simulating plasmonic nanostructures. Unlike the FDTD method, the BEM discretizes the boundary of the nanostructure instead of the entire domain.

In the BEM, Maxwell's equations are transformed into integral equations based on the Green's theorem, which relate the values of the electric and magnetic fields on the boundary of the nanostructure to the current distribution:

$$E(\mathbf{r}) = \frac{1}{4\pi\epsilon} \int_{\partial V} \left(\frac{\mathbf{n} \times J(\mathbf{r}')}{|\mathbf{r}-\mathbf{r}'|} - \frac{J(\mathbf{r}') \times (\mathbf{r}-\mathbf{r}')}{|\mathbf{r}-\mathbf{r}'|^3} \right) \cdot d\mathbf{A}', \quad (10.13)$$

$$H(\mathbf{r}) = \frac{1}{4\pi} \int_{\partial V} \left(\frac{\mathbf{n} \times M(\mathbf{r}')}{|\mathbf{r}-\mathbf{r}'|} - \frac{M(\mathbf{r}') \times (\mathbf{r}-\mathbf{r}')}{|\mathbf{r}-\mathbf{r}'|^3} \right) \cdot d\mathbf{A}', \quad (10.14)$$

where E and H are the electric and magnetic fields, respectively, ϵ is the permittivity of the medium, $\mathbf{n}$ is the outward unit normal vector to the boundary, ∂V is the boundary of the nanostructure, $J(\mathbf{r}')$ is the current density at position $\mathbf{r}'$ on the boundary, $M(\mathbf{r}')$ is the magnetic current density at position $\mathbf{r}'$, and $d\mathbf{A}'$ represents a differential area element on the boundary.

By discretizing the boundary into small surface elements and using appropriate numerical integration schemes, one can solve the integral equations and obtain the electric and magnetic fields inside and outside the nanostructure.

Conclusion

In this chapter, we have explored the simulation of plasmon-enhanced phenomena and their applications in plasmonic nanostructures. By leveraging Maxwell's equations and employing numerical techniques such as the Finite-Difference Time-Domain (FDTD) method and the Boundary Element Method (BEM), researchers can gain valuable insights into the behavior of surface plasmons and design novel plasmonic devices with tailored optical properties. These simulations pave the way for various applications, including biosensing, optical trapping, and enhanced spectroscopy, among others.

In the next chapter, we will delve into the exploration of carrier dynamics in nanostructures. Specifically, we will examine the intricacies of carrier lifetimes, mobility, and recombination processes, shedding light on the underlying mathematical models and simulation techniques.Below is a Python code snippet that implements the essential equations and algorithms discussed in the chapter on plasmonic nanostructures, including the simulation of electromagnetic fields using the Finite-Difference Time-Domain (FDTD) method and the Boundary Element Method (BEM).

Python Code Snippet

The following Python code demonstrates the implementation of FDTD and BEM techniques for simulating plasmonic nanostructures.

```
import numpy as np
import matplotlib.pyplot as plt

def fdtd_simulation(n_steps, n_cells, dt, dx, eps, mu,
    source_position):
    '''
    Finite-Difference Time-Domain (FDTD) simulation.
    :param n_steps: Number of time steps.
    :param n_cells: Number of cells in the discretized space.
    :param dt: Time step size.
    :param dx: Spatial step size.
    :param eps: Permittivity of the medium.
    :param mu: Permeability of the medium.
    :param source_position: Position of the source in the grid.
    :return: Electric and magnetic fields.
    '''
    # Initialize arrays for electric and magnetic fields
    Ez = np.zeros(n_cells)
    Hy = np.zeros(n_cells)

    # Time evolution
    for t in range(n_steps):
        # Update magnetic field
        for i in range(n_cells - 1):
            Hy[i] += dt / (mu * dx) * (Ez[i + 1] - Ez[i])

        # Add a source at the designated position
        Ez[source_position] += np.sin(2 * np.pi * 1e9 * t * dt)  #
            Frequency of 1 GHz

        # Update electric field
        for i in range(1, n_cells):
            Ez[i] += dt / (eps * dx) * (Hy[i - 1] - Hy[i])

    return Ez, Hy

# Boundary Element Method is more complex and handles only
# the surface integral; thus, a simplified approach is shown
def bem_simulation(x, y, eps, current_density):
    '''
    Boundary Element Method (BEM) simulation.
    :param x: x-coordinates of the boundary points.
    :param y: y-coordinates of the boundary points.
    :param eps: Permittivity of the medium.
    :param current_density: Current density on the boundary.
```

```python
    :return: Electric field calculated on the boundary.
    '''
    n = len(x)
    E = np.zeros(n)

    for i in range(n):
        for j in range(n):
            # Calculate the distance
            r = np.sqrt((x[i] - x[j])**2 + (y[i] - y[j])**2)
            if r != 0:
                E[i] += 1 / (4 * np.pi * eps) * current_density[j] /
                ↪     r

    return E

# Simulation parameters
n_steps = 1000
n_cells = 100
dt = 1e-12    # Time step in seconds
dx = 1e-9     # Spatial step in meters
eps = 8.854e-12    # Permittivity of free space
mu = 1.256637e-6    # Permeability of free space
source_position = 50    # Position of the source in grid

# Run FDTD simulation
Ez, Hy = fdtd_simulation(n_steps, n_cells, dt, dx, eps, mu,
↪    source_position)

# Example boundary points for BEM
x_boundary = np.array([1, 2, 3, 4])
y_boundary = np.array([1, 1, 1, 1])
current_density = np.array([1e-6, 2e-6, 1.5e-6, 3e-6])    # Example
↪    current densities

# Run BEM simulation
E_boundary = bem_simulation(x_boundary, y_boundary, eps,
↪    current_density)

# Plotting the results
plt.figure(figsize=(10, 5))
plt.subplot(1, 2, 1)
plt.title('FDTD Simulation: Electric Field (Ez)')
plt.plot(Ez)
plt.xlabel('Cell Index')
plt.ylabel('Electric Field (Ez)')

plt.subplot(1, 2, 2)
plt.title('BEM Simulation: Electric Field on Boundary')
plt.plot(x_boundary, E_boundary, 'ro-')
plt.xlabel('X (Boundary Points)')
plt.ylabel('Electric Field (E)')

plt.tight_layout()
```

```
plt.show()
```

This code defines two main functions:

- `fdtd_simulation` performs the Finite-Difference Time-Domain simulation of electric and magnetic fields over a discretized space.
- `bem_simulation` executes a simplified representation of the Boundary Element Method to calculate the electric field on the boundary based on a given current density.

The scripts utilize NumPy for numerical computations and Matplotlib for visualizing the results of both simulations. The FDTD function models the propagation of electromagnetic waves, while the BEM function provides insight into the electric field along defined boundary points, illustrating the core functionality associated with modeling plasmonic nanostructures.

Chapter 11

Carrier Dynamics in Nanostructures

In this chapter, we provide an overview of the carrier dynamics in nanostructures, focusing on key aspects such as carrier lifetimes, mobility, and recombination processes. Understanding these fundamental processes is crucial for the design and optimization of nanoscale electronic and optoelectronic devices. We delve into the mathematics and physics underlying carrier dynamics, discussing the relevant equations and models used to describe these phenomena.

Introduction

Carrier dynamics play a vital role in the performance of nanostructured semiconductor materials. It encompasses the behavior of charge carriers, including electrons and holes, as they traverse through the nanostructure and interact with various scattering mechanisms. The understanding of carrier dynamics allows us to engineer the properties of these materials for a wide range of applications, including solar cells, transistors, and light-emitting diodes.

In this section, we provide an introduction to carrier dynamics and outline the key parameters that govern the behavior of carriers in nanostructures. We discuss the concept of carrier lifetimes, mobility, and recombination processes, laying the foundation for a deeper exploration of these phenomena.

Carrier Lifetimes

Carrier lifetimes, denoted as τ_c for electrons and τ_v for holes, represent the average time an excess charge carrier resides in a given energy state before recombining or undergoing other scattering events. These lifetimes are influenced by various mechanisms, such as radiative and non-radiative recombination, phonon scattering, and carrier-impurity interactions.

The dynamics of carrier lifetimes can be described using the following differential equations:

$$\frac{dN_c}{dt} = -\frac{N_c}{\tau_c} + G_c, \qquad (11.1)$$

$$\frac{dN_v}{dt} = -\frac{N_v}{\tau_v} + G_v, \qquad (11.2)$$

where N_c and N_v are the concentrations of excess charge carriers (electrons and holes, respectively), G_c and G_v are the generation rates of the respective carriers, and τ_c and τ_v represent their respective lifetimes. These equations describe a balance between carrier generation and recombination processes.

Carrier Mobility

Carrier mobility, denoted as μ_c for electrons and μ_v for holes, characterizes the ease with which charge carriers move through a material under the influence of an electric field. It is a measure of the carrier's average drift velocity per unit electric field.

The mobility of carriers can be quantified using the Einstein relation:

$$\mu_{c,v} = \frac{q\mu_{c,v}}{k_B T}, \qquad (11.3)$$

where q is the elementary charge, $\mu_{c,v}$ represents the carrier mobility, k_B is the Boltzmann constant, and T is the temperature of the system. This relationship connects the carrier mobility with their diffusion coefficient.

Recombination Processes

Recombination processes refer to the annihilation of excess charge carriers, leading to a decrease in the carrier concentration within the material. These processes can occur through several mechanisms, including radiative recombination, non-radiative recombination, Shockley-Read-Hall (SRH) recombination, Auger recombination, and surface recombination.

The rate of recombination can be expressed using the Shockley-Read-Hall model:

$$R_{\text{SRH}} = \frac{n_i^2 - np}{t_n t_p} + \frac{p_i^2 - np}{t_p t_n}, \tag{11.4}$$

where n and p represent the electron and hole concentrations, respectively, n_i and p_i denote the intrinsic carrier concentrations, and t_n and t_p represent the carrier lifetimes for electrons and holes, respectively.

Conclusion

In this chapter, we have provided an overview of carrier dynamics in nanostructures, focusing on carrier lifetimes, mobility, and recombination processes. These fundamental aspects play a crucial role in the performance of nanoscale electronic and optoelectronic devices. The understanding of carrier dynamics enables us to optimize device efficiency and engineer materials with tailored properties. Further chapters will expand on specific aspects and applications of carrier dynamics in nanostructures, providing deeper insights into their behavior and practical implications.

Python Code Snippet

Below is a Python code snippet that calculates carrier lifetimes, mobility, and recombination rates based on the equations discussed in this chapter.

```
def carrier_dynamics(initial_n, initial_p, tau_c, tau_v, G_c, G_v,
     time):
    '''
    Simulate carrier dynamics over a specified time period.
    :param initial_n: Initial concentration of electrons (N_c).
```

```
:param initial_p: Initial concentration of holes (N_v).
:param tau_c: Carrier lifetime for electrons (tau_c).
:param tau_v: Carrier lifetime for holes (tau_v).
:param G_c: Generation rate for electrons (G_c).
:param G_v: Generation rate for holes (G_v).
:param time: Time duration for the simulation.
:return: Final concentrations of electrons and holes after the
↪ specified time.
'''
import numpy as np

# Time array
t = np.linspace(0, time, num=100)
n = np.zeros_like(t)   # Electron concentration
p = np.zeros_like(t)   # Hole concentration

# Initial conditions
n[0] = initial_n
p[0] = initial_p

# Solve differential equations using Euler's method
for i in range(1, len(t)):
    dn_dt = -n[i-1] / tau_c + G_c
    dp_dt = -p[i-1] / tau_v + G_v
    n[i] = n[i-1] + dn_dt * (t[i] - t[i-1])
    p[i] = p[i-1] + dp_dt * (t[i] - t[i-1])

return n, p


def carrier_mobility(temperature):
    '''
    Calculate carrier mobility based on temperature.
    :param temperature: Temperature in Kelvin.
    :return: Carrier mobility in m^2/(V·s).
    '''
    q = 1.602e-19   # Elementary charge in coulombs
    k_B = 1.38e-23  # Boltzmann constant in J/K
    # Example mobility value, can be modified based on specific
    ↪ material
    mobility = (q / (k_B * temperature)) * 1e-4   # Example scaling
    return mobility


def recombination_rate(n, p, n_i, t_n, t_p):
    '''
    Calculate the Shockley-Read-Hall recombination rate.
    :param n: Electron concentration.
    :param p: Hole concentration.
    :param n_i: Intrinsic carrier concentration.
    :param t_n: Electron lifetime.
    :param t_p: Hole lifetime.
    :return: Recombination rate.
```

```
    '''
    R_SRH = ((n_i**2 - n * p) / (t_n * t_p)) + ((n_i**2 - n * p) /
    ↪    (t_p * t_n))
    return R_SRH


# Inputs for the calculations
initial_n = 1e10   # Initial electron concentration (cm^-3)
initial_p = 1e10   # Initial hole concentration (cm^-3)
tau_c = 1e-9   # Electron lifetime (seconds)
tau_v = 1e-9   # Hole lifetime (seconds)
G_c = 1e20   # Electron generation rate (cm^-3/s)
G_v = 1e20   # Hole generation rate (cm^-3/s)
time = 1e-6   # Simulation time (seconds)
temperature = 300   # Temperature in Kelvin
n_i = 1e10   # Intrinsic carrier concentration (cm^-3)
t_n = 1e-9   # Electron lifetime (seconds)
t_p = 1e-9   # Hole lifetime (seconds)

# Calculations
n, p = carrier_dynamics(initial_n, initial_p, tau_c, tau_v, G_c,
↪    G_v, time)
mobility = carrier_mobility(temperature)
R_SRH = recombination_rate(n[-1], p[-1], n_i, t_n, t_p)

# Output results
print("Final Electron Concentration:", n[-1], "cm^-3")
print("Final Hole Concentration:", p[-1], "cm^-3")
print("Carrier Mobility:", mobility, "m^2/(V·s)")
print("Recombination Rate:", R_SRH, "cm^-3/s")
```

This code defines three functions:

- `carrier_dynamics` simulates the concentration of electrons and holes over time given their initial conditions and generation rates.
- `carrier_mobility` calculates the carrier mobility based on temperature.
- `recombination_rate` computes the Shockley-Read-Hall recombination rate given carrier concentrations and lifetimes.

The provided example computes the final concentrations of electrons and holes after a specified time, carrier mobility based on temperature, and the recombination rate for a nanostructured semiconductor, then prints the results.

Chapter 12

Atomistic Simulations

In this chapter, we delve into the fascinating world of atomistic simulations for modeling semiconductor nanostructures at the atomic level. This approach allows us to gain a deep understanding of the structural, electronic, and dynamic properties of materials, providing crucial insights into their behavior and enabling the design of novel nanoscale devices. By employing sophisticated computational techniques and powerful numerical algorithms, atomistic simulations offer a unique avenue for exploring the intricate nature of nanostructures and uncovering their hidden secrets. In this chapter, we will explore the theoretical foundations, numerical methodologies, and practical applications of atomistic simulations in semiconductor research.

Introduction

Atomistic simulations, as the name suggests, involve modeling materials at the atomic scale. At this level of resolution, we can accurately capture the complex interactions between atoms, allowing for a detailed examination of the behavior of nanostructures. By simulating the dynamics and properties of individual atoms, we can obtain valuable insights into the behavior of materials that are not easily observable experimentally.

In this section, we will provide an overview of the theoretical background and numerical techniques used in atomistic simulations. We will discuss the fundamental principles of atomistic modeling and highlight the benefits and limitations of this approach.

Atomic Interactions

At the heart of atomistic simulations lies the description of atomic interactions. The behavior of atoms is governed by interatomic forces, which can be modeled using empirical potentials derived from experimental data or first-principles methods based on quantum mechanics.

In empirical potential methods, such as the widely used classical molecular dynamics (MD) simulations, the interactions between atoms are described by mathematical functions known as force fields. These force fields approximate the potential energy associated with atomic configurations and enable the simulation of the system's dynamics. Different force fields can be employed based on the specific material and properties of interest.

Alternatively, first-principles methods, such as density functional theory (DFT), offer a more rigorous approach by solving the electronic structure problem directly from the fundamental equations of quantum mechanics. These methods are computationally demanding but provide highly accurate descriptions of atomic interactions, making them particularly suitable for studying the electronic and optical properties of semiconductors.

Numerical Algorithms

Simulating the dynamics of thousands or even millions of atoms requires efficient numerical algorithms. Several techniques have been developed to tackle this challenge, such as molecular dynamics, Monte Carlo simulations, and quantum Monte Carlo methods.

Molecular dynamics (MD) simulations numerically solve the classical equations of motion for atoms, allowing the investigation of time-dependent phenomena. The dynamics of atoms are obtained by integrating Newton's equations with respect to time. MD simulations can provide insights into the structural and mechanical properties of nanostructures, including diffusion, phase transitions, and vibrations.

Monte Carlo simulations, on the other hand, focus on the statistical behavior of systems. In these simulations, atoms are randomly moved, and the acceptance or rejection of these moves depends on the change in energy. Monte Carlo methods are particularly useful for studying equilibrium properties, such as thermodynamics and phase diagrams.

Quantum Monte Carlo (QMC) methods combine the concepts of Monte Carlo simulations with the principles of quantum mechanics. These methods provide accurate predictions of atomic and molecular properties by stochastically sampling the many-body wavefunction. QMC methods, however, are computationally demanding and are typically applied to small systems.

Applications

Atomistic simulations have a wide range of applications in semiconductor research. They provide invaluable insights into the structural and electronic properties of nanoscale materials, enabling the design and optimization of advanced devices. Here, we highlight some of the key research areas where atomistic simulations have made significant contributions:

1. **Crystal Growth**: Simulations can elucidate the mechanisms and dynamics of crystal growth, shedding light on the formation of defects, interface structures, and growth kinetics. This knowledge is crucial for controlling the quality and properties of nanoscale materials.

2. **Defects and Doping**: Atomistic simulations allow us to study the effects of defects and dopants on material properties. By understanding the interaction between impurities and host atoms, we can engineer materials with tailored electronic, optical, and magnetic properties.

3. **Mechanical Properties**: Simulations can provide insights into the mechanical behavior of nanostructures, including their elasticity, fracture, and deformation mechanisms. This knowledge is essential for designing nanoscale devices with improved mechanical reliability.

4. **Optoelectronic Properties**: Atomistic simulations enable the investigation of the optical and electronic properties of semiconductors, such as band gaps, carrier dynamics, and light-matter interactions. These insights contribute to the development of efficient nanoscale optoelectronic devices.

5. **Nanoelectronics**: Simulations play a crucial role in the design and optimization of nanoscale electronic devices, such as transistors, diodes, and memory cells. By accurately modeling the device structure and materials, atomistic simulations provide insights into device performance and guide device engineering.

Conclusion

In this chapter, we have explored the field of atomistic simulations for modeling semiconductor nanostructures at the atomic level. By employing sophisticated computational techniques and powerful numerical algorithms, atomistic simulations offer a unique opportunity to unravel the intricacies of nanostructures and gain valuable insights into their behavior. We have discussed the theoretical foundations, numerical methodologies, and practical applications of atomistic simulations in semiconductor research. The ability to model materials at the atomic scale empowers researchers to optimize device performance, design new materials, and unlock the full potential of nanotechnology.

Python Code Snippet

Below is a Python code snippet that demonstrates essential algorithms for atomistic simulations, specifically focusing on classical molecular dynamics (MD) simulations, and includes functions for computing forces and simulating the motion of atoms using the Verlet integration algorithm.

```python
import numpy as np

def calculate_force(positions):
    '''
    Calculate interatomic forces using a Lennard-Jones potential.
    :param positions: Array of atomic positions in nanometers.
    :return: Array of forces acting on each atom in piconewtons.
    '''
    n_atoms = positions.shape[0]
    forces = np.zeros((n_atoms, 3))  # Initialize force array
    epsilon = 1.0  # Depth of the potential well
    sigma = 1.0    # Finite distance at which the potential is zero

    # Calculate forces based on Lennard-Jones potential
    for i in range(n_atoms):
        for j in range(i + 1, n_atoms):
            r_vec = positions[j] - positions[i]
            distance = np.linalg.norm(r_vec)
            # Lennard-Jones force
            force_magnitude = 24 * epsilon * ((2 * (sigma / distance)**13) - (sigma / distance)**7)
            forces[i] += force_magnitude * (r_vec / distance)  # Update force on atom i
```

```python
        forces[j] -= force_magnitude * (r_vec / distance)  #
         ↪ Update force on atom j

    return forces


def verlet_integration(positions, velocities, forces, dt):
    '''
    Perform one step of the Verlet integration method.
    :param positions: Current positions of atoms in nanometers.
    :param velocities: Current velocities of atoms in nanometers per
     ↪ picosecond.
    :param forces: Current forces acting on atoms in piconewtons.
    :param dt: Time step for the integration in picoseconds.
    :return: Updated positions and velocities of atoms.
    '''
    # Update positions
    new_positions = positions + velocities * dt + 0.5 * forces *
     ↪ dt**2

    # Calculate new forces based on updated positions
    new_forces = calculate_force(new_positions)

    # Update velocities
    new_velocities = velocities + 0.5 * (forces + new_forces) * dt

    return new_positions, new_velocities


# Simulation parameters
num_atoms = 5  # Number of atoms
dt = 0.01  # Time step in picoseconds
num_steps = 1000  # Number of time steps

# Initial conditions
np.random.seed(0)  # For reproducibility
initial_positions = np.random.rand(num_atoms, 3)  # Random initial
 ↪ positions
initial_velocities = np.random.rand(num_atoms, 3)  # Random initial
 ↪ velocities

# Perform MD simulation
positions = initial_positions
velocities = initial_velocities

for step in range(num_steps):
    forces = calculate_force(positions)  # Calculate forces
    positions, velocities = verlet_integration(positions,
     ↪ velocities, forces, dt)  # Update positions and velocities

# Output final positions
print("Final Positions (nm):")
```

```
print(positions)
```

This code defines two functions:

- `calculate_force` computes the forces acting on a set of atoms based on the Lennard-Jones potential, utilizing their positions.
- `verlet_integration` implements the Verlet integration algorithm to update atomic positions and velocities over time.

The provided example initializes the positions and velocities of a number of atoms, simulates their motion over a specified time duration using MD methods, and then prints the final positions of the atoms after the simulation.

Chapter 13

Self-Assembly Processes

In this chapter, we delve into the fascinating field of self-assembly processes in nanostructured semiconductors. Self-assembly refers to the spontaneous organization of nanoscale building blocks into complex functional structures or patterns driven by fundamental physical and chemical interactions. Understanding and simulating self-organization in nanostructured semiconductors is of paramount importance for designing and fabricating advanced devices with tailored properties and enhanced functionalities. In this chapter, we will explore mathematical models and computational techniques that enable us to comprehend the underlying principles governing self-assembly and simulate the dynamics of these processes.

Introduction to Self-Assembly

Self-assembly processes are prevalent in nature and have been extensively studied in various disciplines, including materials science, chemistry, and biology. The ability to control and harness self-assembly holds immense potential for the development of novel materials and devices with unique properties. In the context of nanostructured semiconductors, self-assembly plays a crucial role in the formation of ordered nanostructures, such as quantum dots, nanowires, and nanostructured thin films.

1 Fundamental Interactions

At the heart of self-assembly processes lie the fundamental physical and chemical interactions between the constituent building blocks. These interactions can be broadly classified into three categories: entropic, electrostatic, and intermolecular forces.

Entropic Interactions

Entropic interactions arise from the tendency of systems to maximize their entropy or disorder. In self-assembly, the reduction in the configurational entropy of individual building blocks drives the formation of ordered structures. One of the most prominent examples of entropic self-assembly is the formation of block copolymer micelles, where the hydrophobic blocks aggregate to minimize their contact with the surrounding solvent.

Electrostatic Interactions

Electrostatic interactions, governed by Coulomb's law, play a crucial role in self-assembly processes involving charged building blocks. These interactions can be attractive or repulsive depending on the charges and distances involved. Electrostatic self-assembly has been extensively utilized in the formation of colloidal crystals, where charged nanoparticles arrange themselves into ordered arrays.

Intermolecular Forces

Intermolecular forces, including van der Waals forces, hydrogen bonding, and dipole-dipole interactions, are ubiquitous in self-assembly processes. These forces act over short distances and can lead to the formation of specific secondary structures, such as helices or sheets, in molecular self-assembly.

Mathematical Modeling of Self-Assembly

Mathematical modeling provides a powerful tool for understanding and predicting self-assembly processes in nanostructured semiconductors. By formulating appropriate mathematical equations and numerical methods, we can simulate the dynamics of self-assembly and gain insights into the underlying mechanisms. Several mathematical models have been developed to describe self-assembly processes, ranging from simple lattice-based models to more complex continuum models.

1 Lattice Models

Lattice models provide a simplified representation of self-assembly processes by discretizing space into a regular lattice. Each lattice site can be occupied by a building block, and the dynamics of self-assembly are governed by transition rules that dictate the movement or attachment of building blocks. Lattice models can be analyzed using mathematical tools such as Markov chains or cellular automata.

An example of a lattice model is the Ising model, which describes the collective behavior of spin systems. The Ising model has been widely used to study phase transitions and self-assembly phenomena in magnetic materials.

2 Continuum Models

Continuum models aim to capture the dynamics of self-assembly processes at a macroscopic level by describing the evolution of concentration fields. These models rely on partial differential equations (PDEs) to describe the transport and diffusion of building blocks, as well as the interplay between different forces and interactions.

One commonly used continuum model is the Cahn-Hilliard equation, which describes the phase separation and domain coarsening in binary systems. The Cahn-Hilliard equation has been applied to study the formation of ordered nanostructures in self-assembled thin films.

Simulation Techniques

Simulating self-assembly processes requires implementing numerical methods to solve the mathematical models described earlier. These techniques enable us to simulate the time evolution of self-assembling systems and predict the final structures obtained under different conditions. Several simulation methods have been developed, each with its own strengths and limitations.

1 Molecular Dynamics (MD)

Molecular dynamics (MD) simulations are widely employed for studying the dynamics of self-assembly at the atomic or molecular level. MD simulations numerically solve the equations of motion for atoms or molecules, taking into account interatomic or intermolecular interactions. By integrating the equations of motion over time, MD simulations can provide insights into the self-assembly kinetics and the resulting structures.

2 Monte Carlo (MC)

Monte Carlo (MC) simulations are statistical simulation methods that enable us to explore the configuration space of self-assembling systems. In MC simulations, building blocks are randomly moved or exchanged according to certain acceptance criteria based on the system's energy. MC simulations can provide information about the equilibrium thermodynamic properties and phase transitions of self-assembling systems.

3 Brownian Dynamics (BD)

Brownian Dynamics (BD) simulations capture the effect of thermal fluctuations on self-assembly processes. BD simulations combine the deterministic dynamics of MD simulations with random forces reflecting the thermal environment. By including the effects of stochastic noise, BD simulations can provide insights into the role of thermal fluctuations in self-assembly phenomena.

4 Phase Field Methods

Phase field methods are a class of numerical techniques that allow us to simulate the dynamics of self-assembly processes by tracking the evolution of order parameters or concentration fields. In phase field methods, the interface between different phases or domains is represented using a phase field variable, and its evolution is governed by coupled PDEs. These methods can capture the complex dynamics of self-assembly, including phase separation, domain coarsening, and pattern formation.

Conclusion

In this chapter, we have explored the fascinating world of self-assembly processes in nanostructured semiconductors. We have discussed the fundamental interactions driving self-assembly and the mathematical models used to describe these processes. Furthermore, we have presented various simulation techniques, including molecular dynamics, Monte Carlo, Brownian dynamics, and phase field methods, that enable us to study self-assembly at different length and time scales. By studying and simulating self-organization in nanostructured semiconductors, we can unravel the mechanisms governing these processes and pave the way for the design and fabrication of advanced nanoscale materials and devices.

Python Code Snippet

Below is a Python code snippet that simulates self-assembly processes in nanostructured semiconductors using the mathematical models discussed in this chapter. The code includes implementations for lattice models via the Ising model, continuum models using the Cahn-Hilliard equation, and simple simulations using Molecular Dynamics (MD).

```python
import numpy as np
import matplotlib.pyplot as plt

def initialize_lattice(size):
    '''
    Initialize a square lattice with random spin values (-1 or 1).
    :param size: Size of the lattice (e.g., 50 means 50x50).
    :return: Initialized lattice.
    '''
    lattice = np.random.choice([-1, 1], size=(size, size))
    return lattice

def calculate_energy(lattice):
    '''
    Calculate the total energy of the lattice based on the Ising
      model.
    :param lattice: Lattice configuration.
    :return: Total energy.
    '''
    energy = 0
    for i in range(lattice.shape[0]):
        for j in range(lattice.shape[1]):
            # Sum the interactions with nearest neighbors
```

```python
            energy -= lattice[i, j] * (lattice[(i + 1) %
                lattice.shape[0], j] +
                                       lattice[i, (j + 1) %
                                           lattice.shape[1]])
    return energy

def update_lattice(lattice, temperature):
    '''
    Update the lattice configuration using Monte Carlo method.
    :param lattice: Current lattice configuration.
    :param temperature: Temperature for the simulation.
    :return: Updated lattice.
    '''
    size = lattice.shape[0]
    for _ in range(size * size):  # Attempt to update each spin
        i, j = np.random.randint(0, size, size=2)
        # Calculate the change in energy
        dE = 2 * lattice[i, j] * (lattice[(i + 1) % size, j] +
                                  lattice[(i - 1) % size, j] +
                                  lattice[i, (j + 1) % size] +
                                  lattice[i, (j - 1) % size])

        # Decide to accept or reject the spin flip
        if dE < 0 or np.random.rand() < np.exp(-dE / temperature):
            lattice[i, j] *= -1  # Flip the spin
    return lattice

def simulate_ising_model(size, temperature, steps):
    '''
    Run a simple Ising model simulation.
    :param size: Size of the lattice.
    :param temperature: Temperature for the simulation.
    :param steps: Number of steps to simulate.
    :return: Array of energies over time.
    '''
    lattice = initialize_lattice(size)
    energies = np.zeros(steps)
    for step in range(steps):
        update_lattice(lattice, temperature)
        energies[step] = calculate_energy(lattice)
    return energies

def plot_energy(energies):
    '''
    Plot the energy over time.
    :param energies: Array of energy values to plot.
    '''
    plt.figure(figsize=(10, 5))
    plt.plot(energies, label='Energy over Time')
    plt.xlabel('Steps')
    plt.ylabel('Energy')
    plt.title('Ising Model Energy Simulation')
    plt.legend()
```

```
    plt.show()

# Simulation parameters
lattice_size = 50  # Size of the lattice (50x50)
temperature = 2.0  # Temperature
steps = 1000  # Number of steps to simulate

# Run simulation
energies = simulate_ising_model(lattice_size, temperature, steps)

# Plot the results
plot_energy(energies)
```

This code defines several functions:

- `initialize_lattice` creates a random lattice for the Ising model.
- `calculate_energy` computes the total energy based on the spin interactions.
- `update_lattice` employs a Monte Carlo method to update the spin configuration based on temperature.
- `simulate_ising_model` runs the Ising model simulation for a specified number of steps.
- `plot_energy` visualizes the energy of the system over the course of the simulation.

The provided example simulates the Ising model for a square lattice and plots the energy over time, illustrating the dynamics of the self-assembly process in nanostructured semiconductors.

Chapter 14

Strain Engineering in Nanostructures

In this chapter, we explore the techniques and effects of strain in modulating properties in nanoscale semiconductors. Strain engineering has emerged as a powerful tool to tailor the electronic, optical, and mechanical properties of nanostructured materials through the application of external mechanical stresses or lattice mismatches. Understanding the underlying mathematical models and computational methods is crucial for the design and optimization of strained nanostructures with desired functionalities. In this chapter, we delve into the mathematical framework and techniques used in strain engineering, providing expert insight into its effects on various properties in nanostructured semiconductors.

Introduction to Strain Engineering

Strain engineering involves the deliberate modification of the lattice structure and atomic positions in nanostructured semiconductors to induce strain or stress. By controlling the strain distribution, we can effectively manipulate the electronic band structure, carrier mobility, and optical properties of the materials. The ability to engineer strain at the nanoscale opens up new possibilities for device optimization and the development of novel functionalities.

1 Strain and Elasticity

Strain refers to the deformation of a material due to external forces or lattice distortions. It is quantified by the relative change in the material's dimensions and is typically represented by the strain tensor. In nanostructured semiconductors, strain can be induced through various techniques, such as substrate patterning, epitaxial growth, or mechanical bending.

Elasticity is the property of a material to return to its original shape after the removal of external forces. In strained nanostructures, the response to elastic deformation is governed by elastic moduli, which describe the material's resistance to deformation under stress. Understanding the elastic properties of nanostructured semiconductors is crucial for predicting their mechanical stability and designing strain-engineered devices.

2 Effects of Strain in Nanostructures

Strain engineering can have profound effects on the electronic, optical, and mechanical properties of nanostructured semiconductors. The specific impact of strain depends on the material's band structure, crystal symmetry, and strain distribution. Here, we provide an overview of some key effects of strain in nanostructures:

Band Structure Modification

Strain affects the electronic band structure of nanostructured semiconductors by altering the energy levels and band gaps. Depending on the strain type (compressive or tensile) and direction, the band structure can exhibit shifts in energy bands, modifications in band curvatures, or even bandgap engineering. This control over the band structure enables the design of nanostructures with tailored electronic properties, including enhanced carrier mobility and tunable bandgaps.

Carrier Mobility Enhancement

Strain engineering can improve the carrier mobility in nanostructured semiconductors by modifying the effective mass of charge carriers. Strain-induced changes in the band structure can result in variations in the effective mass of both electrons and holes, impacting their transport properties. By selectively inducing strain in specific regions of a nanostructure, one can enhance carrier mobility and achieve superior device performance.

Optical Property Tuning

Strain has a significant impact on the optical properties of nanostructured semiconductors. By modulating the band structure and wavefunction overlap, strain can influence the absorption and emission spectra, excitonic properties, and light-matter interactions in nanostructures. Strain engineering enables the tuning of photoluminescence, absorption coefficients, and even the creation of novel optoelectronic devices.

Mechanical and Thermal Effects

Strain in nanostructures can also affect their mechanical and thermal properties. The presence of strain alters the elastic constants, hardness, and thermal conductivities of materials, making them mechanically flexible or enhancing their thermal conductance. These properties are crucial for the design of strain-engineered devices, such as flexible electronics or high-efficiency thermoelectric materials.

Mathematical Modeling of Strain in Nanostructures

Mathematical modeling provides a theoretical framework for understanding and predicting the effects of strain in nanostructured semiconductors. A variety of mathematical models have been developed to describe the mechanical response, strain distribution, and resulting property modifications in strained nanostructures. Here, we explore some of the key mathematical models used in strain engineering:

1 Elastic Continuum Models

Elastic continuum models describe the deformation and mechanical response of materials under strain. These models treat the material as a continuous medium, represented by a set of partial differential equations (PDEs) derived from the theory of elasticity. By solving these PDEs, one can obtain the strain distribution, stress fields, and deformations induced by external stresses or lattice mismatches.

A fundamental equation in elastic continuum models is the linearized theory of elasticity, given by the equation:

$$\sigma_{ij} = C_{ijkl}\epsilon_{kl}, \qquad (14.1)$$

where σ_{ij} represents the stress tensor, C_{ijkl} is the stiffness tensor, and ϵ_{kl} denotes the strain tensor. The stiffness tensor characterizes the material's elastic properties and can be obtained through experiments, atomistic simulations, or density functional theory calculations.

2 Strain-Gradient Elasticity Models

Strain-gradient elasticity models offer a refined description of strain effects in nanostructured materials. These models account for the gradient of strain, which becomes significant at the nanoscale due to the presence of interfaces, defects, or variations in strain distribution. Strain-gradient elasticity models introduce additional terms in the strain energy density functionals to capture the effects of strain gradients on material properties.

The strain-gradient elasticity model can be represented by the following equation:

$$\sigma_{ij} = C_{ijkl}\varepsilon_{kl} + D_{ijk}\varepsilon_{kl,k}, \qquad (14.2)$$

where D_{ijk} is the additional material property capturing the strain gradient effects. Strain-gradient elasticity models provide a more accurate description of strain-induced phenomena at the nanoscale but require additional experimental or computational data to determine the strain-gradient coefficients.

3 Atomistic Simulations

Atomistic simulations, such as molecular dynamics (MD) or density functional theory (DFT), offer detailed insights into the behavior of strained nanostructures at the atomic level. These simulations numerically integrate the equations of motion for atoms or electrons to investigate the mechanical response, electronic structure, and property modifications induced by strain. Atomistic simulations can provide atomistic-scale information about strain distribution, dislocation formation, and stress relaxation, complementing the continuum models described earlier.

Computational Techniques for Strain Engineering

Simulating the effects of strain in nanostructured semiconductors requires the development and implementation of computational techniques. These techniques enable the efficient calculation of strain distributions, mechanical properties, and resulting property modifications in strained nanostructures. Some computational techniques commonly used in strain engineering include:

1 Finite Element Method (FEM)

The finite element method (FEM) is a widely-used numerical technique for solving partial differential equations. FEM discretizes the domain into a finite number of smaller elements to approximate the continuous problem. In the context of strain engineering, FEM can be employed to solve the elastic continuum models discussed earlier. By dividing the structure into finite elements and solving the derived equations, FEM provides a valuable tool for obtaining the strain distribution and stress fields in complex nanostructures.

2 Boundary Element Method (BEM)

The boundary element method (BEM) is another numerical technique used in strain engineering. BEM, also known as the boundary integral equation method, discretizes the domain's boundary rather than the volume to solve the governing equations. BEM is particularly advantageous for problems with infinite domains or boundary-dominated effects. In strain engineering, BEM can be utilized to calculate the strain fields and stress distributions in structures subjected to particular boundary conditions or external mechanical fields.

3 Molecular Dynamics (MD)

Molecular dynamics (MD) simulations, mentioned earlier as a type of atomistic simulation, are particularly effective for investigating the mechanical response and property modifications in strained nanostructures. In MD simulations, the interactions between particles, such as atoms or molecules, are modeled using interatomic potentials. By solving the equations of motion, MD simulations can provide insights into the deformation mechanisms, stress relaxation processes, and resulting structural changes induced by strain.

Conclusion

In this chapter, we have explored the techniques and effects of strain in modulating properties in nanoscale semiconductors. We have discussed the mathematical models, such as elastic continuum models and strain-gradient elasticity models, used to describe the mechanical response of strained nanostructures. Additionally, we have introduced computational techniques, including the finite element method, boundary element method, and molecular dynamics simulations, for analyzing the strain distribution and predicting property modifications. By understanding the underlying mathematical framework and utilizing appropriate computational techniques, researchers and engineers can harness the potential of strain engineering to design and optimize nanostructured semiconductors with tailored properties and improved functionalities.

Python Code Snippet

Below is a Python code snippet that implements the mathematical models discussed in the chapter, including the calculation of stress and strain in nanostructures, as well as the simulation of carrier mobility influenced by strain.

```python
import numpy as np

def calculate_stress(strain, stiffness_tensor):
    """
    Calculate the stress tensor based on strain and material
    stiffness tensor.
    :param strain: Strain tensor (numpy array).
    :param stiffness_tensor: Stiffness tensor (numpy array).
    :return: Stress tensor (numpy array).
    """
    stress = np.tensordot(stiffness_tensor, strain, axes=([1], [0]))
    return stress

def calculate_strain(stress, stiffness_tensor):
    """
    Calculate the strain tensor based on stress and material
    stiffness tensor.
    :param stress: Stress tensor (numpy array).
```

```python
    :param stiffness_tensor: Stiffness tensor (numpy array).
    :return: Strain tensor (numpy array).
    '''
    from numpy.linalg import inv
    strain = np.tensordot(inv(stiffness_tensor), stress, axes=([1],
    ↪    [0]))
    return strain

def calculate_effective_mass(original_mass, strain):
    '''
    Calculate the effective mass of the charge carriers affected by
    ↪    strain.
    :param original_mass: Original effective mass of the carrier
    ↪    (kg).
    :param strain: Applied strain (scalar).
    :return: Effective mass of the carrier (kg).
    '''
    if strain < 0:
        # Compressive strain reduces the effective mass
        effective_mass = original_mass * (1 + strain)
    else:
        # Tensile strain increases the effective mass
        effective_mass = original_mass * (1 - strain)
    return effective_mass

def calculate_carrier_mobility(effective_mass):
    '''
    Calculate the carrier mobility based on effective mass.
    :param effective_mass: Effective mass of the carrier (kg).
    :return: Carrier mobility (m^2/V·s) assuming constant scattering
    ↪    rate.
    '''
    q = 1.6e-19  # Elementary charge in coulombs
    mobility = q / (effective_mass * 1e6)  # Example relation for
    ↪    mobility
    return mobility

# Example Inputs
strain_tensor = np.array([[0.01, 0, 0],
                          [0, 0.01, 0],
                          [0, 0, 0.01]])  # Small strain tensor
stiffness_tensor = np.array([[150e9, 50e9, 0],
                             [50e9, 150e9, 0],
                             [0, 0, 75e9]])  # Stiffness tensor in
                             ↪    Pa (3x3)
original_mass = 9.11e-31  # Mass of an electron in kg

# Calculations
stress_tensor = calculate_stress(strain_tensor, stiffness_tensor)
strain_calculated = calculate_strain(stress_tensor,
↪    stiffness_tensor)
effective_mass = calculate_effective_mass(original_mass,
↪    strain_tensor[0][0])
```

```
carrier_mobility = calculate_carrier_mobility(effective_mass)

# Output results
print("Stress Tensor:\n", stress_tensor)
print("Calculated Strain Tensor:\n", strain_calculated)
print("Effective Mass of Carrier:", effective_mass, "kg")
print("Carrier Mobility:", carrier_mobility, "m^2/V·s")
```

This code defines four functions:

- `calculate_stress` computes the stress tensor from the strain tensor and the material stiffness tensor.
- `calculate_strain` calculates the strain tensor from the stress tensor and the material stiffness tensor.
- `calculate_effective_mass` estimates the effective mass of charge carriers based on the original mass and applied strain.
- `calculate_carrier_mobility` evaluates the carrier mobility based on the effective mass of the charge carriers.

The example provided simulates the calculation of the stress and strain in a nanostructure, along with the effective mass and mobility of charge carriers, then prints the results.

Chapter 15

Nanostructure Fabrication Techniques

Introduction to Nanostructure Fabrication

Nanostructure fabrication techniques play a crucial role in the development of nanoscale devices and applications. Fabrication at the nanoscale presents unique challenges, requiring precise control over material properties, dimensions, and geometries. In this chapter, we explore the modeling challenges and innovations in nanostructure fabrication, providing expert insight into the mathematical and computational methods used to optimize the fabrication process.

1 Importance of Fabrication in Nanotechnology

Fabrication techniques enable the realization of nanostructures with tailored properties, opening up opportunities for novel functionalities and applications. Precise control over fabrication parameters allows for the manipulation of material properties, such as electrical conductivity, optical response, and mechanical strength. Moreover, advancements in fabrication techniques have driven the miniaturization of devices, leading to the development of nanoelectronics, nanophotonics, nanomedicine, and numerous other fields.

2 Modeling Challenges in Nanostructure Fabrication

Modeling nanostructure fabrication presents several challenges due to the complex interplay between material properties, process parameters, and device performance. To effectively model the fabrication process, it is necessary to understand and account for phenomena such as material growth, surface diffusion, crystalline ordering, and defect formation. Additionally, mathematical models must incorporate the effects of external factors, including temperature, pressure, and ambient environment. Addressing these challenges requires the development of advanced mathematical and computational techniques.

Mathematical Modeling of Nanostructure Fabrication

Mathematical modeling provides a theoretical framework for understanding and optimizing nanostructure fabrication processes. By formulating mathematical equations that describe the underlying physics and chemistry, we can gain insights into the fabrication mechanisms and make predictions about device performance. In the context of nanostructure fabrication, several mathematical models and techniques have been developed:

1 Kinetic Monte Carlo Simulations

Kinetic Monte Carlo (KMC) simulations provide a powerful tool for modeling the dynamics of nanostructure growth and surface diffusion during fabrication. KMC simulates the stochastic motion of atoms or molecules on a lattice, allowing for the prediction of growth rates, surface roughness, and defect densities. By incorporating thermodynamic and kinetic factors, KMC simulations can guide the optimization of fabrication parameters and provide insights into the growth mechanisms.

2 Diffusion Models

Diffusion plays a critical role in many nanostructure fabrication processes, such as chemical vapor deposition and epitaxial growth. Diffusion models describe the movement of atoms or molecules across surfaces, capturing the effects of temperature, concentration gradients, and surface interactions. These models, based on Fick's laws of diffusion, enable the prediction of deposition rates, diffusion lengths, and material intermixing during fabrication.

3 Phase Field Models

Phase field models offer a versatile approach for simulating phase transitions, grain growth, and pattern formation in nanostructure fabrication. These models represent material properties and boundaries as continuous fields, allowing for the investigation of complex morphological changes during fabrication. Phase field models can provide insights into the evolution of microstructures, the formation of defects, and the optimization of processing conditions in nanostructure fabrication.

4 Finite Element Analysis

Finite element analysis (FEA) is a widely-used numerical method for simulating the mechanical behavior of nanostructures during fabrication. FEA discretizes the structure into small elements, solving the governing equations to predict stress and deformation distributions. This technique enables the optimization of fabrication parameters, the design of support structures, and the prediction of structural integrity during fabrication processes such as etching, lithography, and deposition.

Computational Techniques in Nanostructure Fabrication

Computational techniques complement mathematical models to facilitate the simulation and optimization of nanostructure fabrication processes. These techniques aim to replicate real-world conditions, account for various physical phenomena, and provide insights into fabrication challenges and innovations. Several computational techniques commonly employed in nanostructure fabrication include:

1 Molecular Dynamics (MD)

Molecular dynamics (MD) simulations are instrumental in understanding the behavior of atoms and molecules during nanostructure fabrication. These simulations numerically integrate the equations of motion to predict atomic trajectories, allowing for the study of nucleation, growth, and defect formation. MD simulations provide insights into energy barriers, reaction rates, and thermodynamic stability during fabrication processes.

2 Quantum Mechanical Methods

Quantum mechanical methods, such as density functional theory (DFT), provide accurate descriptions of atomic and electronic behavior during nanostructure fabrication. DFT calculations allow for the prediction of material properties, such as bonding energies, band structures, and defect energies. These methods are particularly useful for analyzing materials with atomic-scale precision, guiding the design and optimization of nanostructures through controlled fabrication.

3 Computational Fluid Dynamics (CFD)

Computational fluid dynamics (CFD) simulations enable the modeling and analysis of fluid flows, gas-phase reactions, and heat transfer during nanostructure fabrication processes. CFD techniques, based on solving the Navier-Stokes equations, assist in understanding gas or liquid phase transport, mixing, and reactant distributions. By simulating the flow and heat transfer, CFD provides insights into the optimization of reactant delivery and thermal management in fabrication processes.

4 Data-driven Modeling

Data-driven modeling approaches leverage machine learning algorithms to predict and optimize nanostructure fabrication processes. These techniques utilize large datasets from experiments or simulations to develop predictive models and optimize fabrication parameters. Data-driven models can aid in the discovery of new fabrication techniques, the analysis of process-structure-property relationships, and the accelerated design of nanostructures with desired properties.

Conclusion

Nanostructure fabrication techniques are critical for realizing the potential of nanoscale devices and applications. Mathematical modeling and computational techniques provide valuable tools for understanding and optimizing the fabrication processes. Kinetic Monte Carlo simulations, diffusion models, phase field models, and finite element analysis enable insights into the growth, diffusion, and mechanical behavior of nanostructures. Molecular dynamics, quantum mechanical methods, computational fluid dynamics, and data-driven modeling facilitate a deeper understanding of materials and process optimization. By integrating these modeling and computational techniques, researchers can address the challenges and innovate in nanostructure fabrication, expanding the possibilities for nanotechnology and nanoscience.

Python Code Snippet

Below is a Python code snippet that implements key algorithms and equations discussed in this chapter regarding nanostructure fabrication techniques, specifically focusing on Kinetic Monte Carlo (KMC) simulations, diffusion models, phase field models, and finite element analysis.

```python
import numpy as np
import matplotlib.pyplot as plt

def kinetic_monte_carlo_simulation(num_steps, lattice_size):
    '''
    Perform a Kinetic Monte Carlo simulation of a simple lattice
      growth model.
    :param num_steps: Number of simulation steps.
    :param lattice_size: Size of the lattice.
    :return: Lattice showing the growth process.
    '''
    lattice = np.zeros((lattice_size, lattice_size))

    for step in range(num_steps):
        # Randomly choose a site in the lattice
        x, y = np.random.randint(0, lattice_size, size=2)
        # Attempt to "grow" at that site
        if lattice[x, y] == 0:   # Check if the site is empty
            lattice[x, y] = 1    # Fill the site
    return lattice
```

```python
def diffusion_simulation(diffusion_coefficient, time, num_points):
    '''
    Simulate diffusion using a random walk method.
    :param diffusion_coefficient: Diffusion coefficient (m^2/s).
    :param time: Total time of diffusion (s).
    :param num_points: Number of points to simulate.
    :return: X and Y coordinates of the diffusion.
    '''
    dt = 0.01  # Time step
    steps = int(time / dt)
    x = np.zeros(steps)
    y = np.zeros(steps)

    for i in range(1, steps):
        dx = np.sqrt(2 * diffusion_coefficient * dt) *
        ↪ np.random.normal()
        dy = np.sqrt(2 * diffusion_coefficient * dt) *
        ↪ np.random.normal()
        x[i] = x[i-1] + dx
        y[i] = y[i-1] + dy

    plt.figure(figsize=(10, 10))
    plt.plot(x, y)
    plt.title('Diffusion Simulation')
    plt.xlabel('X Position (m)')
    plt.ylabel('Y Position (m)')
    plt.axis('equal')
    plt.grid()
    plt.show()


def phase_field_model(size, num_iterations):
    '''
    Execute a basic phase field model simulation for grain growth.
    :param size: Size of the grid.
    :param num_iterations: Number of iterations for growth.
    :return: Final phase field.
    '''
    field = np.random.rand(size, size)

    for iteration in range(num_iterations):
        for i in range(1, size-1):
            for j in range(1, size-1):
                # Simple rule for phase field evolution
                field[i, j] = (field[i-1, j] + field[i+1, j] +
                ↪ field[i, j-1] + field[i, j+1]) / 4.0

    return field


def finite_element_analysis(size, force):
    '''
```

```
    Simulate the mechanical response of a nanostructure using a
    ↪ simple finite element method.
    :param size: Size of the grid.
    :param force: Applied force (N).
    :return: Displacement field.
    '''
    displacement = np.zeros((size, size))

    # Apply boundary conditions
    displacement[0, :] = force  # Apply force on one edge

    # Assume a simple linear elasticity model
    for i in range(1, size):
        for j in range(1, size):
            displacement[i, j] = (displacement[i-1, j] +
            ↪    displacement[i, j-1]) / 2.0

    plt.imshow(displacement, cmap='viridis')
    plt.colorbar()
    plt.title('Finite Element Analysis Displacement Field')
    plt.xlabel('X Position')
    plt.ylabel('Y Position')
    plt.show()

# Example Usage
lattice_size = 10
num_steps = 50
diffusion_coefficient = 0.1
time = 5.0
num_points = 1000
grid_size = 50
num_iterations = 100
force = 10  # Applied force in N

# Run simulative models
lattice_growth = kinetic_monte_carlo_simulation(num_steps,
↪    lattice_size)
plt.imshow(lattice_growth, cmap='gray')
plt.title('KMC Simulation Result')
plt.show()

diffusion_simulation(diffusion_coefficient, time, num_points)

phase_field_result = phase_field_model(grid_size, num_iterations)
plt.imshow(phase_field_result, cmap='hot')
plt.title('Phase Field Model Result')
plt.colorbar()
plt.show()

finite_element_analysis(grid_size, force)
```

This code defines four functions:

- `kinetic_monte_carlo_simulation` simulates a simple lattice growth model using Kinetic Monte Carlo methods.
- `diffusion_simulation` implements a random walk method to simulate diffusion in a two-dimensional plane.
- `phase_field_model` executes a basic phase field simulation for grain growth using an iterative method.
- `finite_element_analysis` performs a simple finite element method simulation to show how mechanical forces affect a nanostructure.

This example provides a comprehensive simulation suite for exploring core concepts in nanostructure fabrication, allowing for visualization of KMC growth, diffusion phenomena, phase field behavior, and mechanical response to forces.

Chapter 16

Nanostructured Solar Cells

Introduction to Nanostructured Solar Cells

The field of photovoltaics has witnessed significant advancements in recent years, with nanostructured semiconductor materials emerging as promising candidates for solar cell applications. Nanostructured solar cells offer numerous advantages, including enhanced light absorption, increased charge separation efficiency, and reduced recombination losses. In this chapter, we aim to provide a mathematical framework for modeling the photovoltaic mechanisms in nanostructured semiconductor materials. By understanding the fundamental processes governing the operation of these solar cells, we can optimize their performance and guide the design of more efficient devices.

1 Importance of Modeling Photovoltaic Mechanisms

Insightful mathematical models are essential for comprehending the complex photovoltaic processes that occur in nanostructured semiconductor materials. These models help identify key physical parameters, optimize material properties, and predict device performance under various operating conditions. By understanding the intricate interplay between light absorption, charge generation, transport, and collection, we can enhance the design and efficiency of nanostructured solar cells.

Fundamentals of Nanostructured Solar Cells

Nanostructured solar cells are semiconductor devices that absorb light and convert it into electricity through the photovoltaic effect. These devices typically consist of multiple layers of nanostructured materials, each designed to facilitate efficient charge generation and collection. The key components of a nanostructured solar cell include the following:

1 Absorber Layer

The absorber layer is responsible for capturing photons and generating charge carriers. It is typically made of a semiconductor material with a bandgap that matches the solar spectrum to maximize light absorption. In nanostructured solar cells, the absorber layer is often composed of nanowires, quantum dots, or other nanostructures that offer advantages such as enhanced light trapping and increased surface area.

2 Charge Transport Layers

The charge transport layers facilitate the movement of electrons and holes within the solar cell. These layers are typically composed of distinct materials with appropriate energy levels to efficiently transport charge carriers from the absorber layer to the electrodes. In nanostructured solar cells, the design and engineering of charge transport layers play a crucial role in minimizing recombination losses and maximizing charge collection efficiency.

3 Electrodes

The electrodes in a nanostructured solar cell are responsible for collecting the generated charge carriers and enabling their transport to an external circuit. Typically, the device incorporates a transparent conductive oxide (TCO) layer as the front electrode to allow incoming light to reach the absorber layer. The back electrode, which contacts the charge transport layer, is usually made of a metal or metal alloy with high electrical conductivity.

Mathematical Modeling of Nanostructured Solar Cells

Mathematical models guide our understanding of the underlying physics and help optimize the performance of nanostructured solar cells. Through modeling, we can explore the intricate interplay between light absorption, charge generation, transport, and collection within the device. Several mathematical models and techniques are employed in the study of nanostructured solar cells:

1 Optical Modeling

Optical modeling describes the absorption and scattering of light within the absorber layer of a nanostructured solar cell. Maxwell's equations govern the interaction of light with nanoscale structures, and numerical methods such as finite difference time domain (FDTD), rigorous coupled wave analysis (RCWA), and transfer matrix methods (TMM) are commonly used to solve these equations. Optical modeling allows us to optimize the absorption efficiency and light-trapping properties of nanostructured materials.

2 Charge Generation Modeling

Charge generation modeling involves understanding the mechanisms by which absorbed photons generate electron-hole pairs (excitons) within the absorber layer. For nanostructured materials, the spatial distribution of excitons can differ significantly from that of bulk materials due to enhanced light absorption and increased interfacial area. Quantum mechanical calculations, such as density functional theory (DFT) and time-dependent density functional theory (TD-DFT), can provide insights into exciton formation and recombination dynamics.

3 Charge Transport Modeling

Charge transport modeling encompasses the study of electron and hole motion within the nanostructured solar cell. This involves considering phenomena such as carrier mobility, traps, recombination, and the effects of electric fields. Various transport models, including drift-diffusion equations, Marcus theory, and master equation approaches, can provide insights into charge transport phenomena and guide the design of efficient charge transport layers.

4 Device-Level Modeling

Device-level modeling combines optical, charge generation, and charge transport models to simulate the overall performance of a nanostructured solar cell. These models take into account factors such as electrical contacts, interface recombination, series and shunt resistances, and the spatial distribution of carrier concentrations. Device simulations using techniques such as finite element method (FEM) or circuit-based methods can provide valuable insights into the efficiency, voltage, and current characteristics of nanostructured solar cells.

Conclusion

In this chapter, we explored the mathematical modeling of photovoltaic mechanisms in nanostructured semiconductor materials. By understanding the fundamental processes governing the operation of nanostructured solar cells, we can optimize their performance and guide the design of more efficient devices. Optical modeling, charge generation modeling, charge transport modeling, and device-level modeling provide powerful tools for simulating and analyzing nanostructured solar cells. Through the integration of these models, researchers can gain insights into device efficiency, explore novel material designs, and contribute to the advancement of photovoltaic technology.

Python Code Snippet

Below is a Python code snippet that models various photovoltaic mechanisms in nanostructured solar cells. This includes calculations for optical modeling, charge generation, and transport modeling.

```
import numpy as np
import matplotlib.pyplot as plt

def optical_modeling(thickness, wavelength):
    '''
    Calculate the absorption coefficient using Beer-Lambert law for
    ↪ given thickness and wavelength.
    :param thickness: Thickness of the material in meters.
    :param wavelength: Wavelength of light in meters.
    :return: Absorption coefficient in cm^-1.
```

```python
    '''
    absorption_coefficient = 1.0 / wavelength  # Simplified
    ↪  estimation
    return absorption_coefficient * thickness * 100  # Convert to
    ↪  cm^-1

def charge_generation_modeling(photon_energy, bandgap_energy):
    '''
    Calculate the number of electron-hole pairs generated per
    ↪  absorbed photon.
    :param photon_energy: Energy of the absorbed photon in eV.
    :param bandgap_energy: Bandgap energy of the semiconductor in
    ↪  eV.
    :return: Number of electron-hole pairs generated.
    '''
    if photon_energy >= bandgap_energy:
        return 1  # One electron-hole pair per photon absorbed
    else:
        return 0  # No pairs generated if photon energy is below
        ↪  bandgap

def charge_transport_modeling(charge_mobility, electric_field,
↪  time):
    '''
    Calculate the drift velocity of charge carriers using the
    ↪  mobility and electric field.
    :param charge_mobility: Mobility of charge carriers in
    ↪  m^2/(V*s).
    :param electric_field: Electric field strength in V/m.
    :param time: Time in seconds.
    :return: Drift velocity in m/s.
    '''
    drift_velocity = charge_mobility * electric_field * time
    return drift_velocity

# Parameters
material_thickness = 0.0005  # Thickness in meters (500 micrometers)
incident_wavelength = 500e-9  # Wavelength in meters (500 nm)
bandgap_energy = 1.1  # Bandgap energy in eV (Silicon)
photon_energy = 2.0  # Typical photon energy in eV
charge_mobility = 0.14  # Mobility in m^2/(V*s) (Silicon)
electric_field = 1000  # Electric field in V/m
time = 1e-6  # Time in seconds

# Calculations
absorption_coefficient = optical_modeling(material_thickness,
↪  incident_wavelength)
num_of_pairs = charge_generation_modeling(photon_energy,
↪  bandgap_energy)
drift_velocity = charge_transport_modeling(charge_mobility,
↪  electric_field, time)

# Output results
```

```
print("Absorption Coefficient:", absorption_coefficient, "cm^-1")
print("Number of Electron-Hole Pairs Generated:", num_of_pairs)
print("Drift Velocity of Charge Carriers:", drift_velocity, "m/s")
```

This code defines three functions:

- `optical_modeling` calculates the absorption coefficient using a simplified estimation based on the material's thickness and the incident wavelength.
- `charge_generation_modeling` determines the number of electron-hole pairs generated per absorbed photon based on the energy of the photon relative to the semiconductor's bandgap.
- `charge_transport_modeling` computes the drift velocity of charge carriers under the influence of an electric field using the concept of charge mobility.

The provided example executes optical modeling to determine the absorption coefficient, estimates the number of electron-hole pairs, and calculates the drift velocity of charge carriers in a nanostructured solar cell, then prints the results.

Chapter 17

Excitonic Effects in Nanostructures

Excitonic effects play a crucial role in the behavior and properties of nanostructured materials. In this chapter, we explore the intricate nature of excitons and their impact on the optical and electronic properties of nanomaterials. We delve into the fundamental physics of excitons, investigate their formation and dynamics, and discuss their significance in the context of nanostructures. By understanding excitonic effects, we can gain insights into the unique properties and potential applications of these materials.

Excitons: Basic Concepts and Properties

Excitons are quasiparticles that arise from the bound state of an electron and a hole in a solid. In nanostructures, such as quantum dots or quantum wells, the confinement of charge carriers enhances the interaction between electrons and holes, leading to pronounced excitonic effects. To delve into the behavior of excitons, it is essential to understand their basic concepts and properties.

1 Exciton Formation and Binding Energy

The formation of an exciton occurs when an electron is excited from the valence band to the conduction band, leaving behind a hole. The electron and hole then become localized and interact via the Coulomb interaction. The binding energy of an exciton, denoted as

E_b, represents the energy required to break the electron-hole pair and separate them. It is a crucial parameter that determines the stability and properties of excitons in nanostructures.

The binding energy of an exciton can be calculated using the Rydberg formula:

$$E_b = \frac{\mu e^4}{2(4\pi\epsilon_0)^2 \epsilon_r^2 \hbar^2}, \quad (17.1)$$

where μ is the reduced mass of the electron-hole system, e is the elementary charge, ϵ_0 is the vacuum permittivity, ϵ_r is the relative dielectric constant of the material, and $\hbar$ is the reduced Planck's constant.

2 Exciton Size and Spatial Distribution

The spatial distribution of an exciton is determined by the wave functions of the electron and the hole. The size of an exciton, often characterized by the exciton radius a_B, represents the typical distance between the electron and the hole. In nanostructures, the confinement of charge carriers influences the exciton size and can lead to enhanced or reduced interaction between the electron and the hole.

The exciton radius a_B can be estimated using an approximation based on the effective mass approximation and the hydrogen-like model:

$$a_B = \frac{\epsilon_r \hbar^2}{\mu e^2}, \quad (17.2)$$

where μ is the reduced mass of the electron-hole system, ϵ_r is the relative dielectric constant of the material, $\hbar$ is the reduced Planck's constant, and e is the elementary charge.

Excitonic Effects on Nanomaterial Properties

Excitons have a profound impact on the optical, electronic, and transport properties of nanostructured materials. Understanding and harnessing these effects is crucial for tailoring the properties of nanomaterials for various applications. In this section, we explore the influence of excitons on different aspects of nanomaterial properties.

1 Optical Absorption and Emission

Excitons significantly affect the optical properties of nanostructures, including their absorption and emission spectra. The discrete energy levels of excitonic states give rise to distinct absorption peaks, known as excitonic peaks. These peaks result from the discrete allowed transitions between the electron and hole energy levels in the excitonic system.

The absorption coefficient α of a nanostructured material is directly related to the exciton concentration N_X and the absorption cross-section σ:

$$\alpha = \sigma N_X, \tag{17.3}$$

where σ is given by:

$$\sigma = \frac{4\pi^2 e^2}{3 m_e^2 c^3} f(E_{\text{gap}}) E_{\text{gap}}, \tag{17.4}$$

with m_e representing the electron mass, c being the speed of light, and E_{gap} denoting the energy band gap of the material. The function $f(E_{\text{gap}})$ accounts for the electronic structure of the material and is related to the joint density of states.

2 Exciton Dynamics and Luminescence

The dynamics of excitons govern their emission characteristics and play a crucial role in the luminescence properties of nanostructured materials. Excitons can undergo radiative recombination, where the recombination of the electron and hole is accompanied by the emission of a photon. This radiative recombination process gives rise to luminescence, such as photoluminescence (PL) or electroluminescence (EL).

The radiative recombination rate R_{rad} is proportional to the exciton concentration N_X and the radiative recombination lifetime τ_{rad}:

$$R_{\text{rad}} = \frac{N_X}{\tau_{\text{rad}}}. \tag{17.5}$$

The radiative recombination lifetime τ_{rad} is inversely proportional to the spontaneous emission rate Γ_{rad}:

$$\tau_{\text{rad}} = \frac{1}{\Gamma_{\text{rad}}}. \tag{17.6}$$

3 Exciton-Exciton Interactions

In dense excitonic systems, interactions between excitons become important and can lead to phenomena such as excitonic energy transfer or the formation of exciton complexes. These interactions can affect exciton dynamics, energy dissipation, and the spectral properties of nanostructures.

Exciton-exciton interactions are characterized by the exciton-exciton scattering length a_{ee}, which describes the range of the exciton-exciton interaction potential. The exciton-exciton scattering length a_{ee} is related to the exciton radius a_B:

$$a_{ee} = \frac{1}{4\pi n_X a_B^2}, \tag{17.7}$$

where n_X is the exciton concentration.

Conclusion

In this chapter, we explored excitonic effects in nanostructures, focusing on their behavior and impact on nanomaterial properties. We investigated the concept of excitons, their formation, and binding energy. Excitons play a vital role in modifying the optical and electronic properties of nanostructured materials, particularly in terms of absorption, emission, luminescence, and exciton-exciton interactions. Understanding excitonic effects is crucial for designing and optimizing nanomaterials for a wide range of applications, including optoelectronics, photovoltaics, and quantum technologies.

Python Code Snippet

Below is a Python code snippet that implements important formulas and algorithms discussed in the chapter on excitonic effects in nanostructures. This code allows you to calculate the binding energy of an exciton, exciton radius, absorption coefficient, radiative recombination rate, and the exciton-exciton scattering length.

```python
def calculate_binding_energy(mu, epsilon_r):
    '''
    Calculate the binding energy of an exciton using the Rydberg
    formula.
    :param mu: Reduced mass of the electron-hole system in kg.
```

```python
    :param epsilon_r: Relative dielectric constant of the material.
    :return: Binding energy in joules.
    '''
    import scipy.constants as const
    E_b = (mu * (const.e**4)) / (2 * ((4 * const.pi *
        const.epsilon_0)**2 * (epsilon_r**2) * (const.hbar**2)))
    return E_b

def calculate_excitons_radius(mu, epsilon_r):
    '''
    Estimate the exciton radius.
    :param mu: Reduced mass of the electron-hole system in kg.
    :param epsilon_r: Relative dielectric constant of the material.
    :return: Exciton radius in meters.
    '''
    import scipy.constants as const
    a_B = (epsilon_r * const.hbar**2) / (mu * (const.e**2))
    return a_B

def calculate_absorption_coefficient(sigma, N_X):
    '''
    Calculate the absorption coefficient.
    :param sigma: Absorption cross-section in m^2.
    :param N_X: Concentration of excitons in m^-3.
    :return: Absorption coefficient in m^-1.
    '''
    return sigma * N_X

def calculate_radiative_recombination_rate(N_X, tau_rad):
    '''
    Calculate the radiative recombination rate of excitons.
    :param N_X: Concentration of excitons in m^-3.
    :param tau_rad: Radiative recombination lifetime in seconds.
    :return: Radiative recombination rate in s^-1.
    '''
    return N_X / tau_rad

def calculate_excitonic_scattering_length(a_B, n_X):
    '''
    Calculate the exciton-exciton scattering length.
    :param a_B: Exciton radius in meters.
    :param n_X: Exciton concentration in m^-3.
    :return: Scattering length in meters.
    '''
    return 1 / (4 * np.pi * n_X * a_B**2)

# Constants and inputs for calculations
mu_electron_hole = 9.11e-31  # Reduced mass in kg (approximation for
    electron and hole)
epsilon_r_material = 12  # Relative dielectric constant (example
    value)
N_X_concentration = 1e20  # Concentration of excitons in m^-3
    (example value)
```

```
sigma_absorption = 1e-17  # Absorption cross-section in m^2 (example
↪ value)
tau_rad_lifetime = 1e-9  # Radiative lifetime in seconds (example
↪ value)
n_X_concentration = 1e20  # Exciton concentration in m^-3 (example
↪ value)

# Calculations
binding_energy = calculate_binding_energy(mu_electron_hole,
↪ epsilon_r_material)
excitons_radius = calculate_excitons_radius(mu_electron_hole,
↪ epsilon_r_material)
absorption_coefficient =
↪ calculate_absorption_coefficient(sigma_absorption,
↪ N_X_concentration)
radiative_recombination_rate =
↪ calculate_radiative_recombination_rate(N_X_concentration,
↪ tau_rad_lifetime)
excitonic_scattering_length =
↪ calculate_excitonic_scattering_length(excitons_radius,
↪ n_X_concentration)

# Output results
print("Binding Energy:", binding_energy, "Joules")
print("Exciton Radius:", excitons_radius, "meters")
print("Absorption Coefficient:", absorption_coefficient, "m^-1")
print("Radiative Recombination Rate:", radiative_recombination_rate,
↪ "s^-1")
print("Excitonic Scattering Length:", excitonic_scattering_length,
↪ "meters")
```

This code defines five functions:

- calculate_binding_energy computes the binding energy of an exciton based on the reduced mass and dielectric constant.
- calculate_excitons_radius estimates the radius of the exciton using the reduced mass and dielectric constant.
- calculate_absorption_coefficient calculates the absorption coefficient from the absorption cross-section and exciton concentration.
- calculate_radiative_recombination_rate derives the radiative recombination rate based on exciton concentration and lifetime.
- calculate_excitonic_scattering_length determines the exciton-exciton scattering length from the exciton radius and concentration.

The provided script performs these calculations using example input values and prints the results, allowing for a numerical exploration of excitonic effects in nanostructured materials.

Chapter 18

Photoluminescence in Nanostructures

In this chapter, we delve into the fascinating phenomenon of photoluminescence in nanostructures, focusing on the mathematical modeling of light emission and its applications in optoelectronic devices. Photoluminescence refers to the emission of light from a material upon absorption of photons. In nanostructures, such as quantum dots and nanowires, the confinement of charge carriers and the unique energy band structures give rise to distinct photoluminescence properties. By understanding the physics and mathematics behind photoluminescence, we can develop accurate models to describe and predict the emission behavior and optimize device performance.

Principles of Photoluminescence

Photoluminescence occurs when incident photons are absorbed by a material, exciting electrons from the valence band to higher energy levels in the conduction band. Subsequently, the excited electrons recombine with holes, releasing energy in the form of photons. This process can be described mathematically using rate equations that account for the generation, recombination, and emission of carriers in the nanostructure.

1 Rate Equations

The rate equations for photoluminescence can be derived by considering the generation and recombination processes of charge carriers. Let n and p be the electron and hole densities, respectively, and N denote the density of defects or dopants that can generate or recombine charge carriers. The rate equations can be expressed as:

$$\frac{dn}{dt} = G - R - \frac{n}{\tau_n}, \qquad (18.1)$$

$$\frac{dp}{dt} = G - R - \frac{p}{\tau_p}, \qquad (18.2)$$

where G represents the generation rate of charge carriers, R is the recombination rate, τ_n and τ_p denote the electron and hole lifetimes, respectively.

2 Radiative and Non-Radiative Recombination

In the context of photoluminescence, recombination processes can be classified into radiative and non-radiative recombination. Radiative recombination involves the emission of photons, contributing directly to the photoluminescence signal. On the other hand, non-radiative recombination processes result in energy dissipation as heat or other non-radiative channels, reducing the efficiency of light emission.

The rates of radiative (R_{rad}) and non-radiative ($R_{\text{non-rad}}$) recombination can be expressed as:

$$R_{\text{rad}} = B_{\text{rad}} np, \qquad (18.3)$$

$$R_{\text{non-rad}} = B_{\text{non-rad}} np, \qquad (18.4)$$

where B_{rad} and $B_{\text{non-rad}}$ are the radiative and non-radiative recombination constants, respectively. The total recombination rate R is given by:

$$R = R_{\text{rad}} + R_{\text{non-rad}}. \qquad (18.5)$$

Mathematical Models for Photoluminescence

To describe the photoluminescence behavior in nanostructures, mathematical models are developed based on physical principles and experimental data. These models provide insights into the underlying processes and enable the prediction of photoluminescence properties. In this section, we explore two commonly used mathematical models for photoluminescence: the Rate Equation Model and the Detailed Balance Model.

1 The Rate Equation Model

The Rate Equation Model describes the temporal evolution of the carrier densities n and p using rate equations. By solving the rate equations, one can obtain the time-dependent behavior of photoluminescence emission. This model provides a simple approach to analyze the effects of carrier generation, recombination, and relaxation processes on the photoluminescence dynamics.

2 The Detailed Balance Model

The Detailed Balance Model is based on the principle of equilibrium between the absorption and emission processes in the material. This model describes the equilibrium carrier densities at a given temperature and provides a detailed understanding of the photoluminescence spectra. By solving a system of balance equations, the equilibrium distribution of carriers and the corresponding photoluminescence intensity can be calculated.

Applications of Photoluminescence in Optoelectronic Devices

Photoluminescence has a wide range of applications in optoelectronic devices, including light-emitting diodes (LEDs), lasers, and photovoltaic cells. The ability to control and manipulate the emission properties of nanostructures is crucial for device performance optimization. By employing mathematical models, we can design and engineer nanostructures with tailored photoluminescence behavior for specific applications.

1 LEDs and Lasers

In LEDs and lasers, the efficient generation and extraction of light are essential for device operation. By understanding the photoluminescence processes and using mathematical models, we can optimize the structure and material properties of the active regions in these devices, enhancing the light emission efficiency and achieving higher device performance.

2 Photovoltaic Cells

In photovoltaic cells, the ability to convert light energy into electrical energy is crucial. Photoluminescence can provide valuable insights into charge generation, recombination, and transport mechanisms in nanostructured photovoltaic materials. Mathematical models enable the design of efficient photovoltaic materials with enhanced light absorption and charge extraction capabilities.

Conclusion

In this chapter, we explored the mathematical modeling of photoluminescence in nanostructures and its applications in optoelectronic devices. By understanding the principles of photoluminescence and using rate equations and equilibrium models, we can describe and predict the emission behavior in nanostructured materials. Furthermore, the mathematical models discussed in this chapter provide insights into the optimization of optoelectronic devices, such as LEDs, lasers, and photovoltaic cells, by tailoring the photoluminescence properties.

Python Code Snippet

Below is a Python code snippet that implements the important equations and algorithms discussed in this chapter regarding photoluminescence in nanostructures.

```python
import numpy as np
import matplotlib.pyplot as plt

def photoluminescence_model(G, B_rad, B_non_rad, tau_n, tau_p,
     time_span):
    '''
```

```python
    Simulate the photoluminescence behavior over time using rate
    ↪  equations.

    :param G: Generation rate of charge carriers.
    :param B_rad: Radiative recombination constant.
    :param B_non_rad: Non-radiative recombination constant.
    :param tau_n: Electron lifetime.
    :param tau_p: Hole lifetime.
    :param time_span: Time range for simulation.
    :return: Electron density, hole density, and photoluminescence
    ↪  intensity over time.
    '''
    time_steps = len(time_span)
    n = np.zeros(time_steps)
    p = np.zeros(time_steps)
    PL_intensity = np.zeros(time_steps)

    # Initial conditions
    n[0] = p[0] = 0.1  # Initial electron and hole densities

    for t in range(1, time_steps):
        R_rad = B_rad * n[t-1] * p[t-1]
        R_non_rad = B_non_rad * n[t-1] * p[t-1]
        R = R_rad + R_non_rad

        dn_dt = G - R - n[t-1] / tau_n
        dp_dt = G - R - p[t-1] / tau_p

        n[t] = n[t-1] + dn_dt * (time_span[t] - time_span[t-1])
        p[t] = p[t-1] + dp_dt * (time_span[t] - time_span[t-1])

        # Photoluminescence intensity is proportional to the
        ↪  radiative recombination rate
        PL_intensity[t] = R_rad

    return n, p, PL_intensity


# Parameters for the simulation
G = 1e20  # Carrier generation rate (cm^-3 s^-1)
B_rad = 1e-10  # Radiative recombination constant (cm^3/s)
B_non_rad = 5e-10  # Non-radiative recombination constant (cm^3/s)
tau_n = 1e-7  # Electron lifetime (s)
tau_p = 1e-7  # Hole lifetime (s)
time_span = np.linspace(0, 1e-6, 100)  # Time range (0 to 1
↪  microsecond)

# Run the model
n, p, PL_intensity = photoluminescence_model(G, B_rad, B_non_rad,
↪  tau_n, tau_p, time_span)

# Plotting the results
plt.figure(figsize=(12, 8))
```

```python
plt.subplot(3, 1, 1)
plt.plot(time_span, n, label='Electron Density (n)', color='blue')
plt.plot(time_span, p, label='Hole Density (p)', color='red')
plt.title('Carrier Densities Over Time')
plt.xlabel('Time (s)')
plt.ylabel('Density (cm^-3)')
plt.legend()
plt.grid()

plt.subplot(3, 1, 2)
plt.plot(time_span, PL_intensity * 1e8, label='Photoluminescence
    Intensity', color='green')
plt.title('Photoluminescence Intensity Over Time')
plt.xlabel('Time (s)')
plt.ylabel('Intensity (scaled)')
plt.legend()
plt.grid()

plt.tight_layout()
plt.show()
```

This code defines a function:

- **photoluminescence_model** simulates the photoluminescence behavior of nanostructures over time using rate equations based on the concepts covered in this chapter.

The code calculates the evolution of electron and hole densities, as well as the resulting photoluminescence intensity over a specified time range, and then visualizes the results using Matplotlib. This simulation helps in understanding the dynamics of photoluminescence in nanostructured materials.

Chapter 19

Multiphysical Simulations

In this chapter, we explore the field of multiphysical simulations, which involve the integration of electrical, thermal, and mechanical simulations to develop comprehensive models of physical systems. By combining these different physical domains, we can better understand the interactions and coupling effects that occur in complex systems. Multiphysical simulations provide valuable insights into the behavior of materials and devices under a wide range of conditions, allowing for the optimization of performance and design.

Introduction to Multiphysical Simulations

Multiphysical simulations involve the simultaneous solution of governing equations from multiple physical domains, such as electrical, thermal, and mechanical. These domains are intertwined, and neglecting their interactions can lead to inaccurate predictions of system behavior. By considering the coupled nature of these phenomena, we can develop more realistic and comprehensive models.

1 Motivation and Importance

The motivation behind multiphysical simulations lies in addressing the inherent coupling between different physical processes in many real-world applications. For example, in microelectronics, the electrical performance of devices is affected by thermal effects and mechanical stresses. Similarly, in energy systems, thermal management and structural integrity are influenced by electrical and mechanical factors. By integrating these domains, we can gain a deeper understanding of the underlying physics and predict the overall system behavior accurately.

2 Challenges and Considerations

Multiphysical simulations present several challenges that need to be carefully addressed. Some of these challenges include:

- **Numerical challenges**: Solving the coupled governing equations requires advanced numerical techniques to handle complex and nonlinear systems of equations.

- **Modeling challenges**: Accurately capturing the physics and interactions across different physical domains requires comprehensive models that consider material properties, boundary conditions, and realistic geometries.

- **Validation and verification**: Multiphysical simulations need to be validated and verified against experimental data and benchmark problems to ensure their accuracy and reliability.

- **Computational resources**: Multiphysical simulations are computationally demanding due to the need for solving complex systems of equations. Efficient algorithms and high-performance computing resources are often required to handle the computational workload.

Mathematical Formulation of Multiphysical Simulations

The mathematical formulation of multiphysical simulations involves the establishment of coupled equations that describe the physical interactions between different domains. In this section, we discuss the general approach to formulating and solving the coupled equations.

1 Governing Equations

The governing equations for multiphysical simulations depend on the specific domains being considered. These equations can include, but are not limited to, the following:

- **Electrical domain**: The electrical domain is described by classical equations such as Ohm's law, Kirchhoff's laws, and the continuity equation for charge.

- **Thermal domain**: The thermal domain involves the heat transfer equations, such as the heat conduction equation, convective heat transfer equations, and radiation heat transfer equations.

- **Mechanical domain**: The mechanical domain is governed by equations of solid mechanics, including the equations of motion, stress-strain relations, and boundary conditions.

2 Coupling Terms

The coupling between different physical domains is introduced through appropriate coupling terms in the governing equations. These coupling terms represent the interactions and dependencies between the variables of different domains. For example, in the case of thermal-mechanical coupling, the heat transfer equation can be coupled with the equations of solid mechanics through the thermal expansion term.

3 Solution Techniques

Solving coupled equations from multiple domains requires the use of advanced numerical techniques. Some commonly employed techniques include:

- **Sequential solution**: In this approach, the equations from different domains are solved sequentially, updating the variables of one domain while keeping the variables of other domains constant. This process is iterated until convergence is achieved.

- **Operator splitting**: Operator splitting methods divide the coupled equations into simpler subproblems that are solved individually. The solutions from each subproblem are then combined to obtain the overall solution.

- **Monolithic solution**: Monolithic solution techniques solve the coupled system of equations as a single entity, rather than splitting it into subproblems. This approach allows for the simultaneous consideration of interactions and coupling effects, but typically requires more computational resources.

Applications of Multiphysical Simulations

Multiphysical simulations find applications in a wide range of fields and industries. In this section, we highlight a few important areas where multiphysical simulations have been extensively employed.

1 Microelectronics

In microelectronics, multiphysical simulations are used to study the behavior of electronic devices under different operating conditions. The coupling between electrical, thermal, and mechanical domains is crucial in accurately predicting device performance, heat dissipation, and reliability. Multiphysical simulations aid in the design and optimization of microelectronic components such as transistors, interconnects, and integrated circuits.

2 Energy Systems

Multiphysical simulations play a key role in the analysis and design of energy systems, including power generation, transmission, and storage. Understanding the interactions between electrical, thermal, and mechanical aspects is essential for maximizing efficiency and ensuring the integrity of system components. Multiphysical simulations are used to study phenomena such as electrical power losses, thermal management, and structural integrity in energy systems.

3 Biomechanics and Biomedical Engineering

In biomechanics and biomedical engineering, multiphysical simulations are employed to investigate the behavior of biological tissues and organs. The coupling between mechanical, electrical, and fluidic domains allows for the study of phenomena such as tissue deformation, fluid flow, and electrical stimulation. Multiphysical simulations assist in the design and optimization of medical devices and treatments, such as implants, prosthetics, and drug delivery systems.

Conclusion

Multiphysical simulations enable a comprehensive understanding of complex physical systems by combining electrical, thermal, and mechanical simulations. By addressing the coupled nature of different physical domains, multiphysical simulations provide insights into the interactions and coupling effects that occur in various applications. The mathematical formulation of multiphysical simulations involves the establishment of coupled equations and the use of advanced numerical techniques for their solution. Multiphysical simulations find applications in microelectronics, energy systems, biomechanics, and biomedical engineering, among others. Through multiphysical simulations, researchers and engineers can optimize designs, improve performance, and ensure the reliable operation of systems and devices.

Python Code Snippet

Below is a Python code snippet that models the coupled electrical and thermal domains in a multiphysical simulation. This example demonstrates how to calculate electrical current using Ohm's Law, heat conduction via the Fourier's Law, and provide an interface for updating states within a coupled simulation framework.

```
import numpy as np

def calculate_current(voltage, resistance):
    '''
    Calculate electrical current based on Ohm's Law.
    :param voltage: Voltage across the resistor in volts.
    :param resistance: Resistance in ohms.
    :return: Current in amperes.
    '''
    return voltage / resistance

def calculate_heat_flux(thermal_conductivity, area,
    ↪ temperature_gradient):
    '''
    Calculate heat flux using Fourier's Law of heat conduction.
    :param thermal_conductivity: Thermal conductivity of the
    ↪    material in W/(m·K).
```

```
    :param area: Cross-sectional area through which heat is being
    ↪ conducted in square meters.
    :param temperature_gradient: Temperature gradient in K/m.
    :return: Heat flux in W/m².
    '''
    return thermal_conductivity * area * temperature_gradient

def update_temperature(current, resistance, time, specific_heat,
↪ mass):
    '''
    Update temperature based on the electric current and heat
    ↪ generated.
    :param current: Current in amperes.
    :param resistance: Resistance in ohms.
    :param time: Time interval in seconds.
    :param specific_heat: Specific heat capacity of the material in
    ↪ J/(kg·K).
    :param mass: Mass of the material in kg.
    :return: Change in temperature in Kelvin.
    '''
    power = current ** 2 * resistance  # Power (P = I^2 * R)
    heat_added = power * time  # Total heat added over the time
    ↪ interval
    return heat_added / (mass * specific_heat)  # Temperature change
    ↪ (T = Q/m*c)

# Constants
voltage = 10  # Volts
resistance = 5  # Ohms
thermal_conductivity = 200  # W/(m·K)
area = 0.01  # m²
temperature_gradient = 50  # K/m
time_interval = 1  # seconds
specific_heat = 900  # J/(kg·K)
mass = 0.1  # kg

# Calculations
current = calculate_current(voltage, resistance)
heat_flux = calculate_heat_flux(thermal_conductivity, area,
↪ temperature_gradient)
temperature_change = update_temperature(current, resistance,
↪ time_interval, specific_heat, mass)

# Output results
print("Calculated Current:", current, "A")
print("Heat Flux:", heat_flux, "W/m²")
print("Temperature Change:", temperature_change, "K")
```

This code defines three functions:

- calculate_current computes the electric current using Ohm's

Law.

- `calculate_heat_flux` uses Fourier's Law to calculate the heat flux in the thermal domain.

- `update_temperature` updates the temperature based on the heat generated by the electrical current.

The provided code then calculates the current, heat flux, and the change in temperature for the simulated scenario, and prints the results.

Chapter 20

Hybrid Nanostructures

In this chapter, we delve into the fascinating world of hybrid nanostructures, focusing specifically on the modeling of properties and applications of hybrid organic-inorganic nanocomposites. These materials, composed of a combination of inorganic and organic components, exhibit unique and tunable properties that make them highly attractive for various technological applications, including optoelectronics, sensing, energy conversion and storage, and biomedical devices. By employing mathematical and computational modeling techniques, we can gain insights into the behavior of hybrid nanostructures and guide their design and optimization.

Introduction to Hybrid Nanostructures

Hybrid nanostructures refer to composite materials that combine nanoscale organic and inorganic components to synergistically exploit their combined properties. The organic component often consists of organic molecules, polymers, or biomolecules, while the inorganic counterpart includes nanoparticles, nanowires, or thin films. The integration of these different components at the nanoscale leads to enhanced functionalities and novel phenomena that are not present in their individual constituents.

Hybrid nanostructures have gained significant attention due to their tailored properties, which can be precisely controlled by manipulating their composition, size, shape, and interface properties. The ability to tune these properties has opened up avenues for designing materials with desired characteristics for specific applications.

1 Characteristics of Hybrid Nanostructures

Hybrid nanostructures possess several distinct characteristics that make them highly appealing for various applications:

- **Tunable properties**: By adjusting the composition and structure of hybrid nanostructures, it is possible to tune their optical, electrical, mechanical, and thermal properties. This tunability provides flexibility in tailoring materials based on specific application requirements.

- **Synergistic effects**: The combination of organic and inorganic components in hybrid nanostructures leads to synergistic effects, where the properties of the resulting material are significantly enhanced or modified compared to the individual components. This synergistic behavior arises from the strong coupling and interactions between the organic and inorganic components at the nanoscale.

- **Multifunctionality**: Hybrid nanostructures have the potential to exhibit multiple functionalities, such as light emission, energy harvesting, sensing, and biomedical capabilities, within a single material system. This multifunctionality opens up diverse application possibilities and enables the development of highly integrated and compact devices.

- **Versatile fabrication techniques**: Hybrid nanostructures can be synthesized using various fabrication techniques, including chemical synthesis, self-assembly, physical deposition, and lithographic processes. These techniques provide flexibility in controlling the morphology, size, and spatial arrangement of the organic and inorganic components, allowing for the creation of complex nanostructures.

Modeling Approaches for Hybrid Nanostructures

Mathematical and computational modeling approaches play a crucial role in understanding and predicting the behavior of hybrid nanostructures. By formulating appropriate mathematical models and employing efficient computational techniques, we can gain insights into their structure-property relationships and guide their design for specific applications. In this section, we discuss some commonly employed modeling approaches for hybrid nanostructures.

1 Continuum Models

Continuum models are widely used to describe the behavior of hybrid nanostructures at length scales larger than the characteristic sizes of their individual components. These models treat the hybrid material as a homogeneous medium with averaged properties and assume that the microscopic details of the constituent components can be neglected.

The governing equations for continuum models depend on the specific properties and functionalities being studied. For example, the electrical behavior of hybrid nanostructures can be modeled using the classical equations of electromagnetism, such as Maxwell's equations and Ohm's law. Similarly, the mechanical behavior can be described using the equations of linear elasticity or viscoelasticity, depending on the time scale of interest.

Continuum models provide a macroscopic understanding of hybrid nanostructures and can capture their overall behavior, but they may not fully account for the nanoscale effects and interfacial phenomena that are crucial in these materials.

2 Particle-based Models

Particle-based models, such as molecular dynamics (MD) and Monte Carlo (MC) simulations, offer a more atomistic description of hybrid nanostructures. These models consider the individual atoms or molecules as discrete particles and track their positions, velocities, and interactions over time.

MD simulations are particularly useful for studying the structural and dynamical properties of hybrid nanostructures, such as their stability, self-assembly, and thermal behavior. These simulations solve the equations of motion for each particle, taking into account interatomic potentials that govern their interactions.

MC simulations, on the other hand, are valuable for investigating configurational changes in hybrid nanostructures. By randomly sampling different configurations, MC simulations can explore the phase space and provide insights into the thermodynamics and phase transitions of the materials.

Particle-based models provide valuable atomistic-level insights into the behavior of hybrid nanostructures. However, they are computationally expensive and are typically limited to relatively small system sizes and short time scales.

3 Multiscale Models

Hybrid nanostructures often exhibit behaviors that span multiple length and time scales, necessitating the use of multiscale modeling techniques. These models aim to bridge the gap between the atomistic and continuum scales, enabling the study of phenomena that occur at different length and time regimes.

Multiscale models employ a hierarchy of models, where each model captures the behavior of the system at a particular scale. The information from one scale is then passed on to the next scale through appropriate coupling conditions or upscaling/downscaling techniques.

For example, a multiscale model for hybrid nanostructures can involve combining atomistic models, such as MD simulations, with continuum models, such as finite element methods (FEM). This allows for the atomistic details to be captured at the nanoscale while still considering the overall behavior of the material at larger length scales.

Multiscale models provide a powerful tool for understanding the complex behavior of hybrid nanostructures and their interplay across different scales. However, developing and implementing these models require careful consideration of the interfaces between different scales and the transfer of information across these interfaces.

Applications of Hybrid Nanostructures

Hybrid nanostructures find applications in various fields due to their unique properties and functionalities. In this section, we highlight a few key areas where hybrid nanostructures have been extensively investigated for their applications.

1 Optoelectronics

Hybrid nanostructures are being actively explored for optoelectronic devices, such as light-emitting diodes (LEDs), photovoltaic cells (solar cells), and photodetectors. By judiciously combining organic and inorganic components, these devices can exhibit efficient light emission, energy harvesting, and light sensing capabilities.

The modeling of hybrid nanostructures in optoelectronics involves understanding their excitation and relaxation processes, charge transfer dynamics, and light-matter interactions. Additionally, factors such as recombination pathways, exciton diffusion, and charge transport play significant roles in determining device performance.

2 Sensing and Biosensing

Hybrid nanostructures have promising applications in sensing and biosensing due to their high surface-to-volume ratios, chemical variability, and biofunctionalization potential. These materials can be tailored to detect specific analytes, such as chemical species, biomolecules, or disease markers, with enhanced sensitivity and selectivity.

The modeling of hybrid nanostructures for sensing applications involves understanding their surface chemistry, surface plasmon resonance, and charge transfer at interfaces. Moreover, the diffusion and binding of analytes, as well as the signal transduction mechanisms, are crucial aspects that can be investigated through modeling approaches.

3 Energy Conversion and Storage

Hybrid nanostructures have demonstrated great potential in energy conversion and storage devices, including batteries, supercapacitors, and fuel cells. Their tailored properties, such as high specific surface area, tunable bandgaps, and fast charge transfer kinetics, enable improved energy storage and conversion efficiencies.

The modeling of hybrid nanostructures in energy devices involves understanding their electrochemical behavior, ion transport phenomena, and interface reactions. Techniques like electrochemical impedance spectroscopy, charge-discharge cycling simulations, and thermodynamic modeling aid in optimizing device performance, stability, and longevity.

4 Biomedical Devices and Therapeutics

Hybrid nanostructures have garnered significant interest in biomedical applications, including drug delivery, bioimaging, and tissue engineering. The combination of organic and inorganic components provides capabilities for targeted drug delivery, multimodal imaging, and enhanced cellular interactions.

Modeling approaches for hybrid nanostructures in biomedical applications focus on understanding their biodistribution, cellular uptake, drug release kinetics, and toxicity. Additionally, simulations aid in the design and optimization of hybrid nanostructures for tailored therapeutic strategies, including personalized medicine and regenerative therapies.

Conclusion

In this chapter, we have explored the modeling of properties and applications of hybrid organic-inorganic nanostructures. These materials offer unique and tunable properties, stemming from the combination of organic and inorganic components, for a wide range of technological applications. Mathematical and computational modeling techniques play a crucial role in understanding the behavior of hybrid nanostructures and guiding their design and optimization. Continuum models, particle-based models, and multiscale models have emerged as valuable tools for investigating the properties and functionalities of hybrid nanostructures. Moreover, hybrid nanostructures find applications in optoelectronics, sensing, energy conversion and storage, and biomedical devices, providing numerous avenues for further research and development.

Python Code Snippet

Below is a Python code snippet that implements key equations and algorithms related to the modeling of hybrid nanostructures, focusing on the calculation of optical properties, mechanical behavior, and charge transport dynamics.

```python
import numpy as np
import matplotlib.pyplot as plt

# Constants
hbar = 1.0545718e-34   # Reduced Planck's constant in J*s
e = 1.602176634e-19    # Elementary charge in Coulombs
m_e = 9.10938356e-31   # Mass of electron in kg

def exciton_energy(n):
```

```python
    '''
    Calculate the exciton energy level using the Rydberg formula for
    ↪ hydrogen-like systems.
    :param n: Principal quantum number (n=1,2,...).
    :return: Energy of the exciton in Joules.
    '''
    Rydberg = 13.6 * e  # Rydberg constant in Joules
    return Rydberg / n**2

def effective_mass(m_n, m_p):
    '''
    Calculate the effective mass of charge carriers in a hybrid
    ↪ nanostructure.
    :param m_n: Effective mass of electrons in kg.
    :param m_p: Effective mass of holes in kg.
    :return: Effective mass in kg.
    '''
    return (m_n * m_p) / (m_n + m_p)

def carrier_mobility(effective_mass, electric_field):
    '''
    Calculate the carrier mobility in a semiconductor.
    :param effective_mass: Effective mass in kg.
    :param electric_field: Electric field strength in V/m.
    :return: Carrier mobility in m^2/(V*s).
    '''
    return (e / (effective_mass * electric_field))

# Simulation parameters
n_levels = 10  # Calculate for first 10 energy levels for excitons
quantum_numbers = np.arange(1, n_levels + 1)  # Quantum numbers

# Calculate exciton energies
energies = [exciton_energy(n) for n in quantum_numbers]

# Define electron and hole effective masses
m_electron = 0.26 * m_e  # Effective mass of electrons
m_hole = 0.39 * m_e  # Effective mass of holes

# Calculate effective mass
m_effective = effective_mass(m_electron, m_hole)

# Define an electric field
electric_field = 1e5  # Electric field in V/m

# Calculate mobility
mobility = carrier_mobility(m_effective, electric_field)

# Output the values
print("Exciton Energies (Joules):", energies)
print("Effective Mass (kg):", m_effective)
print("Carrier Mobility (m^2/(V*s)):", mobility)
```

```
# Plotting the exciton energy levels
plt.plot(quantum_numbers, energies)
plt.title("Exciton Energy Levels")
plt.xlabel("Principal Quantum Number (n)")
plt.ylabel("Exciton Energy (J)")
plt.grid()
plt.show()
```

This code defines three functions:

- `excitons_energy` calculates the energy levels of excitons using the Rydberg formula.
- `effective_mass` computes the effective mass of charge carriers in a hybrid nanostructure.
- `carrier_mobility` calculates the carrier mobility based on effective mass and electric field.

The provided example computes the exciton energies, effective mass, and carrier mobility, then plots the exciton energy levels against the principal quantum numbers. The results are printed to the console for further analysis.

Chapter 21

Graphene and 2D Materials

In this chapter, we delve into the advanced modeling techniques for graphene and other two-dimensional (2D) materials with unique properties. Graphene, a monolayer of carbon atoms arranged in a hexagonal lattice, has garnered significant attention due to its exceptional mechanical, electrical, thermal, and optical properties. The field of 2D materials has now expanded beyond graphene to include various other materials, such as transition metal dichalcogenides (TMDs) and black phosphorus, each with its own set of intriguing characteristics.

Introduction to Graphene and 2D Materials

Graphene, being the most extensively studied 2D material, has remarkable properties that have attracted considerable interest from both the scientific and engineering communities. Its electronic band structure exhibits linear dispersion near the Dirac points, leading to high carrier mobility and exceptional electronic conductivity. Furthermore, its mechanical properties are unparalleled, with a tensile strength orders of magnitude higher than steel. In addition to graphene, a diverse family of 2D materials has emerged, each with its own unique properties that can be tailored for specific applications.

1 Crystal Structure of Graphene

Graphene possesses a hexagonal crystal structure, with each carbon atom covalently bonded to its three neighboring carbon atoms. The lattice can be visualized as a monolayer of carbon atoms arranged in a honeycomb pattern. This hexagonal lattice structure gives rise to many of the exceptional properties of graphene.

2 Electronic Band Structure of Graphene

The electronic band structure of graphene is described by the -bands resulting from the overlap of p-orbitals of carbon atoms. The low-energy electronic states near the Dirac points, where the valence and conduction bands touch, lead to unique electronic properties, such as massless Dirac fermions and linear band dispersion.

3 Other 2D Materials

Apart from graphene, there is a wide range of 2D materials that exhibit intriguing properties. Some of the notable ones include transition metal dichalcogenides (TMDs), such as molybdenum disulfide (MoS_2) and tungsten diselenide (WSe_2), hexagonal boron nitride (h-BN), and black phosphorus. Each of these materials has distinct characteristics due to their different crystal structures and chemical compositions.

Modeling Approaches for Graphene and 2D Materials

Mathematical and computational modeling techniques play a vital role in understanding and predicting the properties of graphene and other 2D materials. By formulating appropriate theoretical models, combined with efficient computational algorithms, we can gain insights into their behavior and guide experimental design and optimization.

1 Electronic Band Structure Calculations

To understand the electronic properties of graphene and 2D materials, it is essential to calculate their band structures. Various theoretical methods, such as density functional theory (DFT), tight-binding approximation, and many-body perturbation theory, can be employed for this purpose. These methods provide valuable information about the energy levels, bandgaps, and electronic states of the materials.

2 Transport Modeling

Transport phenomena in graphene and 2D materials are of significant interest for device applications. Transport models allow for the understanding of charge carrier motion, conductivity, and current-voltage characteristics. Approaches such as the Boltzmann transport equation, semiclassical transport theory, and quantum transport simulations provide insights into charge transport mechanisms, carrier mobility, and quantum transport effects.

3 Strain and Defect Engineering

Strain engineering, the deliberate application of strain to modify material properties, has gained attention in the context of graphene and 2D materials. Theoretical models incorporating strain effects allow for the prediction of modified band structures, changes in electronic properties, and the emergence of pseudo-magnetic fields. Furthermore, defect engineering, the intentional introduction of defects into materials, can be modeled to understand their impact on the mechanical, electrical, and optical properties of 2D materials.

4 Optical and Photonic Properties

The optical properties of graphene and 2D materials are crucial for applications in optoelectronics and photonics. Modeling techniques can be employed to investigate light-matter interactions, such as absorption, emission, and scattering processes. Moreover, the optical response can be tailored through external stimuli, such as electric and magnetic fields or strain, allowing for the design of tunable photonic devices.

5 Heterostructures and van der Waals Interactions

Heterostructures composed of different layered 2D materials offer unique opportunities to engineer material properties. Modeling techniques that incorporate van der Waals interactions between the layers enable the understanding of interlayer coupling effects, band alignment, and the emergence of novel electronic and optical phenomena. Theoretical investigations can guide the design of heterostructures with tailored functionalities.

Conclusion

In this chapter, we have explored advanced modeling techniques for graphene and other 2D materials. Their exceptional mechanical, electrical, thermal, and optical properties make them highly desirable for a wide range of applications. Mathematical and computational modeling approaches, such as electronic band structure calculations, transport modeling, strain and defect engineering, optical property modeling, and investigations of heterostructures, provide crucial insights into the behavior and properties of these materials. By harnessing these modeling techniques, we can effectively design and optimize graphene and 2D materials for various technological applications.Below is a comprehensive Python code snippet encapsulating essential equations and algorithms relevant to the modeling of graphene and other 2D materials. This code covers electronic band structure calculations using a simple tight-binding model, transport modeling under the Boltzmann transport equation, strain effects, and optical property modeling.

Python Code Snippet

Below is a Python code snippet that implements the important equations and algorithms discussed in this chapter related to modeling graphene and 2D materials.

```python
import numpy as np
import matplotlib.pyplot as plt

def tight_binding_model(k, t):
    '''
    Calculate the energy bands for graphene using the tight-binding
    ↪ model.
    :param k: Wave vector (numpy array).
    :param t: Hopping parameter (eV).
    :return: Energy values (numpy array).
    '''
    return np.array([-2 * t * (np.cos(k[0]) + np.cos(k[1]))])
```

```python
def boltzmann_transport(mu, T, E, k_B):
    '''
    Calculate the conductivity using the Boltzmann transport
    ↪ equation.
    :param mu: Carrier mobility (m^2/Vs).
    :param T: Temperature (K).
    :param E: Electric field (V/m).
    :param k_B: Boltzmann constant (J/K).
    :return: Conductivity (S/m).
    '''
    return mu * (2 * k_B * T) ** (1/2) * E


def strain_effects(original_property, strain):
    '''
    Model the effect of strain on a property.
    :param original_property: Original value of the property.
    :param strain: Applied strain (decimal).
    :return: Modified property.
    '''
    # Assuming linear dependency on strain
    return original_property * (1 + strain)


def optical_absorption_coefficient(E, alpha_0, photon_energy):
    '''
    Calculate the optical absorption coefficient for a 2D material.
    :param E: Energy (eV).
    :param alpha_0: Constant related to absorption (cm^-1).
    :param photon_energy: Photon energy (eV).
    :return: Absorption coefficient (cm^-1).
    '''
    return alpha_0 * (E - photon_energy) ** 2


# Constants
t = 2.7  # Hopping parameter for graphene in eV
k_B = 8.617e-5  # Boltzmann constant in eV/K
mu = 2000e-4  # Carrier mobility in m^2/Vs
T = 300  # Temperature in K
E = 1e3  # Electric field in V/m
strain = 0.05  # 5% applied strain
alpha_0 = 10  # Absorption coefficient parameter in cm^-1
photon_energy = 1.5  # Photon energy in eV

# Wave vector for k-points
k_values = np.linspace(-np.pi, np.pi, 100)
energy_values = tight_binding_model(k_values, t)

# Calculations
conductivity = boltzmann_transport(mu, T, E, k_B)
modified_property = strain_effects(100, strain)  # Original property
↪ = 100
```

```
absorption_coefficient = optical_absorption_coefficient(2, alpha_0,
↪   photon_energy)

# Output results
print("Energy values from Tight Binding Model:", energy_values)
print("Conductivity:", conductivity, "S/m")
print("Modified property after strain:", modified_property)
print("Optical Absorption Coefficient:", absorption_coefficient,
↪   "cm^-1")

# Plotting the energy bands
plt.plot(k_values, energy_values, label='Tight Binding Model')
plt.title('Energy Bands of Graphene')
plt.xlabel('Wave Vector (k)')
plt.ylabel('Energy (eV)')
plt.axhline(0, color='gray', linestyle='--')
plt.axvline(0, color='gray', linestyle='--')
plt.grid()
plt.legend()
plt.show()
```

This code defines four functions:

- tight_binding_model calculates the energy bands of graphene based on the tight-binding model.

- boltzmann_transport computes the conductivity using the Boltzmann transport equation.

- strain_effects models the effect of applied strain on a material property.

- optical_absorption_coefficient calculates the optical absorption coefficient for a 2D material.

The provided example computes the energy bands from a tight-binding model, evaluates conductivity, assesses the effect of strain on a property, and calculates the optical absorption coefficient for a given photon energy. The results are printed, and a plot of energy bands is displayed for visualization.

Chapter 22

Quantum Transport in Nanostructures

In this chapter, we delve into a detailed study of quantum transport phenomena in reduced dimensions. Transport properties in nanostructures play a crucial role in determining their electronic and optoelectronic behavior. The unique characteristics of quantum confinement and reduced dimensionality give rise to a variety of intriguing phenomena, such as ballistic transport, quantized conductance, and the emergence of energy bands. Theoretical models and computational methods have been developed to understand and predict these transport phenomena accurately. In this chapter, we will explore the mathematical foundations and theoretical frameworks for modeling quantum transport in nanostructures, providing a comprehensive understanding of this fascinating field.

Theoretical Framework for Quantum Transport

The theoretical framework for describing quantum transport in nanostructures is based on the principles of quantum mechanics. We consider electron transport through a nanostructure, where the behavior of electrons is governed by wave functions and their associated probabilities. The Schrödinger equation provides the foundation for understanding the evolution of the electron wave function.

1 Mathematical Description of the Schrödinger Equation

The time-independent Schrödinger equation in one dimension is given by:

$$\hat{H}\psi(x) = E\psi(x)$$

where $\hat{H}$ is the Hamiltonian operator, $\psi(x)$ is the wave function, and E is the energy of the system. In nanostructures, the Hamiltonian operator incorporates the effects of confinement, potential barriers, and scattering mechanisms. To solve the Schrödinger equation for a given nanostructure, suitable boundary conditions must be imposed, taking into account the geometry and boundary conditions of the system.

2 Boundary Conditions for Nanostructures

The choice of boundary conditions is critical in solving the Schrödinger equation for nanostructures. For one-dimensional systems, commonly encountered boundary conditions include:

Infinite Potential Barrier: $\psi(0) = \psi(L) = 0$
Periodic Boundary Conditions: $\psi(0) = \psi(L)$, $\psi'(0) = \psi'(L)$
Dirichlet Boundary Conditions: $\psi(0) = \psi(L) = \text{constant}$

The selection of the appropriate boundary conditions depends on the specific characteristics of the nanostructure under investigation.

Transport Parameters and Quantum Mechanics

In quantum transport, various parameters are used to characterize the electron dynamics and transport properties in nanostructures. These parameters provide insights into the conductance, current, and other transport phenomena.

1 Transmission Coefficient

The transmission coefficient describes the probability of an electron to transmit through the nanostructure when incident from one side. It is defined as the ratio of the flux of transmitted electrons to that of incident electrons. This coefficient is commonly calculated using scattering theory and quantum mechanical wave functions.

2 Landauer-Büttiker Formula

The Landauer-Büttiker formula relates the conductance of a nanostructure to the transmission coefficient and the physical properties of the system. It is given by:

$$G = \frac{e^2}{h} \sum_{n,m} T_{nm}$$

where G is the conductance, e is the elementary charge, h is Planck's constant, T_{nm} is the transmission coefficient from mode n to mode m. The summation is performed over all relevant electronic modes.

3 Density of States

The density of states characterizes the number of available energy states per unit energy interval. In nanostructures, the density of states exhibits quantization due to quantum confinement. The energy spacing between quantized energy levels depends on the size and shape of the nanostructure. The density of states is fundamental in determining carrier transport, optical absorption, and electronic transitions.

Computational Methods for Quantum Transport

Numerical methods are essential for solving the Schrödinger equation and calculating transport properties. Various computational techniques have been developed to study quantum transport in nanostructures, each with its strengths and limitations.

1 Transfer Matrix Method

The transfer matrix method is a powerful approach for numerically solving the Schrödinger equation and calculating transmission coefficients in nanostructures with potential barriers. By dividing the nanostructure into discrete slices and incorporating the appropriate boundary conditions, the method allows for the determination of transmission probabilities and conductance.

2 Green's Function Method

The Green's function method provides a comprehensive framework for describing quantum transport. It involves calculating the Green's function, which relates the dynamics of an electron to the properties of the nanostructure and the scattering mechanisms. The Green's function can be used to determine transmission coefficients, conductance, and other transport parameters.

3 Non-Equilibrium Green's Function Formalism

The non-equilibrium Green's function formalism extends the Green's function method to systems driven out of equilibrium by external biases or voltages. It enables the calculation of current-voltage characteristics and device performance, accounting for the interactions between electrons and the scattering potentials.

Conclusion

This chapter has outlined the theoretical and mathematical foundations for understanding quantum transport in nanostructures. The Schrödinger equation serves as the central equation governing electron dynamics, and appropriate boundary conditions provide the necessary constraints for solving the equation. Transport parameters, such as the transmission coefficient and the Landauer-Büttiker formula, offer insights into the conductance and transmission properties. To explore the behavior of electrons and determine transport characteristics, various computational methods, including the transfer matrix method, Green's function method, and non-equilibrium Green's function formalism, have been developed. By employing these theoretical frameworks and computational techniques, researchers can provide valuable insights into the transport behavior of electrons in nanostructures, fueling further advancements in nanoelectronics and nanodevices.

Python Code Snippet

Below is a Python code snippet that calculates the transmission coefficient, conductance using the Landauer-Büttiker formula, and performs multiple quantum transport simulations based on the principles mentioned in this chapter.

```python
import numpy as np
import matplotlib.pyplot as plt

def calculate_transmission_coefficient(incident_wave_func,
     transmitted_wave_func):
    '''
    Calculate the transmission coefficient for a quantum system.
    :param incident_wave_func: Wave function of the incident wave.
    :param transmitted_wave_func: Wave function of the transmitted
     wave.
    :return: Transmission coefficient (T).
    '''
    incident_flux = np.abs(incident_wave_func)**2
    transmitted_flux = np.abs(transmitted_wave_func)**2
    T = transmitted_flux / incident_flux
    return T

def landauer_buttiker_conductance(transmission_coefficients):
    '''
    Calculate the conductance using the Landauer-Büttiker formula.
    :param transmission_coefficients: List of transmission
     coefficients for different modes.
    :return: Total conductance (G).
    '''
    e = 1.602176634e-19  # Elementary charge in coulombs
    h = 6.62607015e-34   # Planck's constant in Js
    G = (e**2 / h) * np.sum(transmission_coefficients)
    return G

def density_of_states(E, width):
    '''
    Calculate the density of states for a 1D system.
    :param E: Energy values.
    :param width: Width of the nanostructure.
    :return: Density of states corresponding to energy values.
    '''
    hbar = 1.0545718e-34   # Reduced Planck's constant in Js
    m_e = 9.10938356e-31   # Electron mass in kg
    DOS = (width / (2 * np.pi)) * np.sqrt((2 * m_e) / (hbar**2 * E))
    return DOS

# Example parameters
width = 1e-9  # Width of the nanostructure in meters
```

```
energy_levels = np.linspace(1e-19, 1e-18, 100)  # Energy levels from
↪ 0 to 1e-18 Joules

# Calculate the density of states
dos = density_of_states(energy_levels, width)

# Example wave functions (for simulation purposes)
incident_wave_func = np.exp(1j * k * x)  # Placeholder for incident
↪ wave function
transmitted_wave_func = np.exp(1j * k_prime * x)  # Placeholder for
↪ transmitted wave function
k = 1e6  # Arbitrary wave number for incident
k_prime = 0.8e6  # Arbitrary wave number for transmitted

# Calculate the transmission coefficient
T = calculate_transmission_coefficient(incident_wave_func,
↪ transmitted_wave_func)

# Calculate conductance
G = landauer_buttiker_conductance([T])

# Output results
print("Transmission Coefficient:", T)
print("Conductance:", G, "S (Siemens)")

# Plotting the Density of States
plt.figure(figsize=(8, 5))
plt.plot(energy_levels, dos, label='Density of States')
plt.xlabel('Energy (Joules)')
plt.ylabel('Density of States (states/Joule/m)')
plt.title('Density of States in a 1D Nanostructure')
plt.legend()
plt.grid()
plt.show()
```

This code defines three functions:

- `calculate_transmission_coefficient` computes the transmission coefficient based on the incident and transmitted wave functions.
- `landauer_buttiker_conductance` applies the Landauer-Büttiker formula to derive the total conductance from the transmission coefficients.
- `density_of_states` estimates the density of states for a one-dimensional system given the energy levels and width of the structure.

The provided example calculates the transmission coefficient from given wave functions, computes conductance, and plots the density of states over a specified energy range, then prints the results.

Chapter 23

Localized Surface Plasmon Resonance (LSPR)

In this chapter, we delve into the fascinating world of Localized Surface Plasmon Resonance (LSPR) and its potential applications in sensing. LSPR is a phenomenon that occurs when free electrons collectively oscillate in response to incident electromagnetic waves at the surface of metallic nanoparticles. The resonant behavior of these surface plasmons induces strong light-matter interactions, leading to unique optical properties that can be exploited for advanced sensing applications. In this chapter, we will focus on the mathematical modeling of LSPR and explore its potential in various sensing platforms.

Introduction to Localized Surface Plasmon Resonance

Localized Surface Plasmon Resonance (LSPR) refers to the collective oscillation of free electrons on the surface of metallic nanoparticles in response to incident electromagnetic waves. This resonance phenomenon occurs due to the confinement and localization of surface plasmons, which are the quantized charge density waves associated with these free electrons.

LSPR is primarily observed in noble metals such as gold and silver, where the interaction between the incident light and the metal nanoparticles leads to the excitation of surface plasmons. The resonant frequency at which this excitation occurs depends on various factors, including the size, shape, and dielectric environment of the nanoparticles. These factors can be finely tuned to achieve specific sensing capabilities.

Mathematical Modeling of LSPR

Mathematical models play a crucial role in understanding and predicting the optical properties of LSPR. The behavior of surface plasmons can be described using the Maxwell's equations coupled with appropriate boundary conditions and material properties.

1 Maxwell's Equations for LSPR

In the presence of LSPR, the coupled Maxwell's equations describe the interaction between the incident electromagnetic field and the surface plasmons. In the frequency domain, these equations can be written as:

$$\nabla \cdot \mathbf{E} = \frac{\rho}{\varepsilon_0} \quad \text{(Gauss's law)}$$

$$\nabla \cdot \mathbf{B} = 0 \quad \text{(Magnetic divergence-free condition)}$$

$$\nabla \times \mathbf{E} = -\frac{\partial \mathbf{B}}{\partial t} \quad \text{(Faraday's law)}$$

$$\nabla \times \mathbf{B} = \mu_0 \mathbf{J} + \mu_0 \varepsilon_0 \frac{\partial \mathbf{E}}{\partial t} \quad \text{(Ampere-Maxwell's law)}$$

where $\mathbf{E}$ and $\mathbf{B}$ are the electric and magnetic fields, respectively, ρ is the charge density, ε_0 is the vacuum permittivity, μ_0 is the vacuum permeability, and $\mathbf{J}$ is the current density.

2 Boundary Conditions and Material Properties

To accurately model LSPR, suitable boundary conditions and material properties must be considered. At the interface between the nanoparticle and the surrounding medium, the continuity of tangential electric and magnetic fields should be satisfied. These boundary conditions, combined with the appropriate material properties such as the dielectric function, determine the behavior of the surface plasmons.

3 Plasmon Resonance Condition

The resonance condition for LSPR is determined by the matching of the incident electromagnetic waves with the dispersion relation of the surface plasmons. The dispersion relation relates the frequency of the incident electromagnetic wave to the wave vector of the surface plasmons, given by:

$$\text{Re}\left[\varepsilon_m(\omega)\right] = -\frac{\varepsilon_p}{\varepsilon_{\text{medium}}}$$

where $\varepsilon_m(\omega)$ is the dielectric function of the metal nanoparticles, ε_p is the permittivity of the metal, and $\varepsilon_{\text{medium}}$ is the permittivity of the surrounding medium.

Applications of LSPR in Sensing

LSPR-based sensing platforms have gained significant attention due to their exceptional sensitivity and label-free detection capabilities. By strategically designing the geometry and composition of the metal nanoparticles, LSPR can be tailored to interact with specific analytes, enabling the detection and quantification of a wide range of target molecules.

1 Nanoparticle-based LSPR Sensors

Nanoparticle-based LSPR sensors typically involve the immobilization of selective recognition elements, such as antibodies or DNA probes, onto the surface of metal nanoparticles. The binding of target analytes to these recognition elements induces a change in the local refractive index, thereby modulating the LSPR signal. These changes can be analyzed to quantify the concentration of the target analyte accurately.

2 LSPR in Surface Enhanced Raman Spectroscopy (SERS)

Surface Enhanced Raman Spectroscopy (SERS) is a powerful technique that combines the enhanced Raman scattering effect with LSPR. Metal nanoparticles, typically gold or silver, are used to amplify the Raman scattering signals of analyte molecules. The excitation of LSPR enhances the local electric field around the nanoparticles, leading to a significant increase in Raman signals.

This enables highly sensitive molecular detection and fingerprinting.

3 LSPR-based Biosensors

LSPR-based biosensors offer unique advantages in terms of sensitivity, multiplexing capability, and real-time monitoring. These biosensors utilize the selective binding of biomolecules onto the surface of metal nanoparticles to induce changes in the LSPR signals. By monitoring these changes, the presence and concentration of specific biomarkers or pathogens can be determined, making LSPR-based biosensors promising for various applications in biomedical diagnostics and environmental monitoring.

Conclusion

Localized Surface Plasmon Resonance (LSPR) is a fascinating phenomenon that has shown great potential in sensing applications. The mathematical modeling of LSPR enables a comprehensive understanding of the optical resonances occurring on the surface of metallic nanoparticles. By accurately describing the behavior of surface plasmons using Maxwell's equations and appropriate boundary conditions, researchers can predict and optimize the optical properties of LSPR. This understanding has paved the way for the development of LSPR-based sensing platforms, including nanoparticle-based sensors, Surface Enhanced Raman Spectroscopy, and biosensors. These applications leverage the exceptional sensitivity and unique light-matter interactions induced by LSPR, offering promising solutions for a broad range of sensing and detection challenges.

Python Code Snippet

Below is a Python code snippet that calculates the resonance condition for Localized Surface Plasmon Resonance (LSPR) and simulates the electric field distribution around metallic nanoparticles.

```
import numpy as np
import matplotlib.pyplot as plt

def dielectric_function(frequency):
    '''
```

```python
    Calculate the dielectric function of a metal based on its
    ↪   frequency.
    :param frequency: Frequency of the incident electromagnetic wave
    ↪   in hertz.
    :return: Dielectric function of the metal.
    '''
    # Example for gold
    epsilon_infinity = 1.0
    plasma_frequency = 1.37e15  # in Hz
    damping_coefficient = 0.08e15  # in Hz
    epsilon = epsilon_infinity - (plasma_frequency**2) /
    ↪   (frequency**2 + 1j * damping_coefficient * frequency)
    return epsilon

def resonant_frequency(epsilon_dispersion, epsilon_medium):
    '''
    Calculate the resonant frequency for the condition of localized
    ↪   surface plasmons.
    :param epsilon_dispersion: Dielectric function of the metal
    ↪   nanoparticles.
    :param epsilon_medium: Permittivity of the surrounding medium.
    :return: Resonant frequency for LSPR.
    '''
    resonant_freq = np.sqrt(-epsilon_dispersion.real /
    ↪   epsilon_medium)
    return resonant_freq

def electric_field_distribution(size, frequency, num_points=100):
    '''
    Simulate the electric field distribution around a nanoparticle.
    :param size: Size of the nanoparticle.
    :param frequency: Frequency of the incident electromagnetic
    ↪   wave.
    :param num_points: Number of points in the simulation grid.
    :return: x, y grid and corresponding electric field values.
    '''
    x = np.linspace(-3*size, 3*size, num_points)
    y = np.linspace(-3*size, 3*size, num_points)
    X, Y = np.meshgrid(x, y)

    epsilon_metal = dielectric_function(frequency)
    field_strength = np.where(X**2 + Y**2 <= size**2,
                              ↪   np.cos(resonant_frequency(epsilon_metal,
                              ↪   1.0)),
                              0)

    return X, Y, field_strength

# Inputs for the calculations
frequency = 5e14  # Frequency in Hz (e.g., 500 THz)
size = 50e-9  # Size of nanoparticle in meters (e.g., 50 nm)
```

```
# Calculate the dielectric function
epsilon_metal = dielectric_function(frequency)
print("Dielectric Function of Metal:", epsilon_metal)

# Calculate the resonance frequency
epsilon_medium = 1.0  # Permittivity of air/vacuum
res_freq = resonant_frequency(epsilon_metal, epsilon_medium)
print("Resonant Frequency for LSPR:", res_freq)

# Simulate the electric field distribution
X, Y, e_field = electric_field_distribution(size, frequency)

# Plot the electric field distribution
plt.figure(figsize=(8, 6))
plt.contourf(X, Y, e_field, levels=50, cmap='RdYlBu')
plt.colorbar(label='Electric Field Strength')
plt.title('Electric Field Distribution around a Nanoparticle')
plt.xlabel('x (m)')
plt.ylabel('y (m)')
plt.axis('equal')
plt.show()
```

This code defines three functions:

- `dielectric_function` computes the dielectric function of a metal based on the frequency of the incident light.
- `resonant_frequency` calculates the resonant frequency for LSPR according to the dielectric conditions.
- `electric_field_distribution` simulates the electric field distribution around a nanoparticle given its size and frequency.

The provided example computes the dielectric function and resonant frequency for a nanoparticle and visualizes the resulting electric field distribution around it.

Chapter 24

Surface Chemistry in Nanostructures

In this chapter, we will explore the fascinating world of surface chemistry in nanostructures and its role in tailoring their properties. Surface interactions and modifications play a crucial role in determining the physical, chemical, and electronic characteristics of nanostructured materials. By understanding and controlling these surface processes, researchers can enhance the performance and functionality of various nanostructured devices. We will delve into the mathematical modeling and analysis of surface chemistry phenomena to provide a comprehensive understanding of this field.

Surface Adsorption and Desorption Processes

Surface chemistry in nanostructures involves the adsorption and desorption of atoms, molecules, or ions on the surface. This interplay between the material and its surrounding environment significantly influences the material's properties and behavior. Understanding these processes is of utmost importance in tailoring the functionality and performance of nanostructured materials.

The adsorption process can be described by Langmuir's adsorption isotherm, which relates the surface coverage (θ) to the concentration of adsorbates (C). It follows the equation:

$$\theta = \frac{kC}{1+kC}$$

where k is the adsorption constant. This equation quantifies the equilibrium distribution of adsorbates between the surface and the gas or liquid phase.

Similarly, the desorption process can be described by a desorption constant, indicating the likelihood of desorption from the material's surface. The desorption process is typically influenced by factors such as temperature and the interaction strength between the adsorbate and the surface.

Surface Modifications and Functionalization

Surface modifications and functionalization techniques are employed to modify the chemical and physical properties of nanostructures. These techniques can be used to enhance surface reactivity, control surface charge, introduce specific functional groups, or create well-defined surface patterns. By tailoring the surface properties, researchers can achieve desired functionalities and applications.

One common method of surface modification is the grafting of functional groups onto the material's surface. For instance, silanization is a widely used technique for covalently attaching organic molecules or polymers to the surface of silicon-based nanostructures. This process involves the reaction of surface hydroxyl groups with silane coupling agents, resulting in the formation of strong chemical bonds.

Mathematically describing these surface modifications requires the consideration of reaction kinetics, thermodynamics, and the chemical nature of the adsorbates and the substrate. The Langmuir-Hinshelwood and Eley-Rideal models are commonly used to describe the kinetics of surface reactions.

Surface Interface Models

To gain a deeper understanding of surface chemistry in nanostructures, various mathematical models and simulations are employed. These models help elucidate the fundamental surface processes and predict the resulting material properties.

The Density Functional Theory (DFT) is an extensively used method for modeling nanoscale surface systems. DFT allows for accurate calculations of electronic structure, energetics, and chemical reactivity. By solving the Kohn-Sham equations, which describe the electron density, DFT provides insights into adsorption energies, reaction barriers, and surface stability.

In addition to DFT, other methods such as molecular dynamics simulations and Monte Carlo simulations are employed to study surface phenomena, including adsorption, diffusion, and surface reactions. These simulations enable the exploration of complex surface processes and provide valuable insights into the behavior of nanostructured materials.

Surface Catalysis and Reactivity

Surface chemistry plays a vital role in catalysis, where the presence of a catalyst enhances the reaction rate without being consumed. Nanostructured surfaces offer unique advantages in catalysis due to their high surface area and reactivity. Understanding the underlying surface processes is crucial for designing efficient catalysts.

Surface catalysis can involve various mechanisms, including adsorption of reactants, chemical reactions on the surface, and desorption of products. The Langmuir-Hinshelwood and Eley-Rideal models mentioned earlier are commonly used to describe these catalytic processes mathematically.

Surface reactivity can be quantified by reaction rates and activation energies. The Arrhenius equation provides a mathematical relationship between the reaction rate and temperature:

$$k = A \exp\left(-\frac{E_a}{RT}\right)$$

where k is the reaction rate, A is the pre-exponential factor, E_a is the activation energy, R is the gas constant, and T is the temperature.

By understanding and manipulating surface reactivity, researchers can design catalysts with optimal activity and selectivity for desired chemical reactions.

Conclusion

In this chapter, we have explored the fascinating field of surface chemistry in nanostructures. Surface interactions and modifications are vital in tailoring the properties and functionality of nanostructured materials. By understanding the adsorption, desorption, and surface reaction processes, researchers can enhance the performance of nanostructured devices. Through mathematical modeling and simulations, we can gain valuable insights into surface phenomena and guide the development of advanced nanostructures with tailored surface properties. The understanding of surface chemistry paves the way for innovations in various fields, including catalysis, sensing, and energy conversion.

Python Code Snippet

Below is a Python code snippet that implements the key equations and algorithms discussed in the chapter on surface chemistry in nanostructures.

```python
import numpy as np
import matplotlib.pyplot as plt

def langmuir_adsorption(C, k):
    '''
    Calculate surface coverage based on Langmuir's adsorption
     isotherm.
    :param C: Concentration of adsorbates.
    :param k: Adsorption constant.
    :return: Surface coverage (theta).
    '''
    theta = (k * C) / (1 + k * C)
    return theta

def arrhenius_equation(A, E_a, T):
    '''
    Calculate the reaction rate using the Arrhenius equation.
    :param A: Pre-exponential factor.
    :param E_a: Activation energy (in Joules).
    :param T: Temperature (in Kelvin).
    :return: Reaction rate (k).
    '''
    R = 8.314  # Gas constant in J/(mol*K)
    k = A * np.exp(-E_a / (R * T))
    return k
```

```python
def plot_adsorption_isotherm(k):
    '''
    Plot the Langmuir adsorption isotherm.
    :param k: Adsorption constant.
    '''
    C = np.linspace(0, 10, 100)  # Concentration range
    theta = langmuir_adsorption(C, k)

    plt.figure(figsize=(8, 5))
    plt.plot(C, theta, label='Langmuir Isotherm', color='blue')
    plt.title('Langmuir Adsorption Isotherm')
    plt.xlabel('Concentration (C)')
    plt.ylabel('Surface Coverage ()')
    plt.axhline(1, color='red', linestyle='--', label='Max
    ↪ Coverage')
    plt.legend()
    plt.grid()
    plt.show()

# Parameters for calculations
adsorption_constant = 0.5  # Adsorption constant (example value)
A = 1e7  # Pre-exponential factor (example value)
E_a = 7500  # Activation energy in Joules (example value)
temperature = 298  # Temperature in Kelvin (25°C)

# Calculations
C_values = np.linspace(0, 10, 100)  # Concentration range for
↪ adsorption
theta_values = langmuir_adsorption(C_values, adsorption_constant)
reaction_rate = arrhenius_equation(A, E_a, temperature)

# Output results
print("Surface Coverage () for varying concentrations:",
↪ theta_values)
print("Reaction Rate (k) at T =", temperature, "K:", reaction_rate)

# Plotting the adsorption isotherm
plot_adsorption_isotherm(adsorption_constant)
```

This code defines three functions:

- `langmuir_adsorption` calculates the surface coverage based on Langmuir's adsorption isotherm using the concentration of adsorbates and the adsorption constant.
- `arrhenius_equation` computes the reaction rate based on the Arrhenius equation, given the pre-exponential factor, activation energy, and temperature.
- `plot_adsorption_isotherm` plots the Langmuir adsorption isotherm to visually represent the relationship between concentration and surface coverage.

The provided example calculates surface coverage for varying concentrations, computes the reaction rate, and generates a plot of the adsorption isotherm, then outputs the results.

Chapter 25

Nanomagnetism

In this chapter, we delve into the captivating realm of nanomagnetism, focusing on the understanding of magnetic behaviors in nanoscale semiconductors and their technological implications. Nanoscale semiconductors refer to semiconductor materials with dimensions on the order of nanometers, where quantum effects become significant and dictate unique magnetic properties. We will explore the mathematical description of these properties and the underlying theoretical frameworks.

Magnetic Moments and Spins

The starting point for understanding nanomagnetism lies in comprehending the behavior of magnetic moments and spins. At the nanoscale, individual atoms can possess magnetic moments due to the alignment of electron spins within the atom. These magnetic moments can interact with each other, leading to collective magnetic behavior in the material.

A crucial concept in nanomagnetism is the spin, a fundamental quantum property of particles. Spins can be in a parallel or antiparallel orientation, giving rise to different magnetic configurations. The Bloch vector, or magnetization vector, is commonly used to represent these orientations quantitatively. It describes the average magnetic moment per unit volume and provides insights into the magnetic properties of nanoscale semiconductors.

Exchange Interactions

Exchange interactions play a pivotal role in determining the magnetic properties of nanoscale semiconductors. These interactions arise from the exchange of electrons between neighboring atoms, creating a coupling between their spins. The strength and nature of exchange interactions greatly influence the magnetic behavior of nanomaterials.

The Heisenberg model is widely used to describe exchange interactions in nanoscale semiconductors. It considers the interaction energy between pairs of spins and can be expressed mathematically as:

$$H = -J \sum_{i \neq j} \mathbf{S}_i \cdot \mathbf{S}_j$$

where H is the exchange energy, J is the exchange integral, $\mathbf{S}_i$ and $\mathbf{S}_j$ are the spin vectors of the ith and jth atoms, and the sum is taken over all pairs of interacting spins.

The exchange interaction can result in various magnetic ordering phenomena, such as ferromagnetism, antiferromagnetism, or ferrimagnetism, depending on the alignment of spins. These magnetic orders can exhibit unique properties at the nanoscale, leading to enhanced functionalities and applications.

Magnetic Anisotropy

Magnetic anisotropy arises from the dependence of a material's magnetic properties on the spatial orientation of its magnetic moments. It characterizes the preferred direction of magnetization in the material. Understanding and controlling magnetic anisotropy is crucial for tailoring the magnetic properties of nanoscale semiconductors.

Various types of magnetic anisotropy can exist, including shape anisotropy, crystal anisotropy, and magnetoelastic anisotropy. Shape anisotropy arises from the shape of the material, crystal anisotropy depends on the crystal structure, and magnetoelastic anisotropy arises from the interaction between the magnetic moments and the surrounding lattice.

Mathematically, magnetic anisotropy can be described using energy terms. For instance, the uniaxial anisotropy energy can be expressed as:

$$E_{\text{anisotropy}} = K\sin^2(\theta)$$

where $E_{\text{anisotropy}}$ is the anisotropy energy, K is the anisotropy constant, and θ is the angle between the magnetization vector and the preferred direction.

Dynamics of Nanomagnetism

Understanding the dynamics of nanomagnetism is crucial for the design and optimization of magnetic devices and technologies. The dynamics are governed by the competition between different energy terms, including exchange interactions, anisotropy, external fields, and thermal effects.

Micromagnetics is a powerful theoretical framework that describes the dynamics of magnetization in nanoscale systems. It incorporates the Landau-Lifshitz-Gilbert (LLG) equation, which provides a mathematical description of the time evolution of the magnetization vector **M**:

$$\frac{d\mathbf{M}}{dt} = -\gamma \mathbf{M} \times \mathbf{H}_{\text{eff}} + \alpha \mathbf{M} \times \frac{d\mathbf{M}}{dt}$$

where γ is the gyromagnetic ratio, $\mathbf{H}_{\text{eff}}$ is the effective magnetic field, and α is the dimensionless damping parameter.

The LLG equation captures the precession of the magnetization vector around the effective magnetic field and the damping of this precession due to dissipation effects. Numerical methods and simulations are often employed to solve the LLG equation and study the dynamics of nanomagnetism.

Technological Implications

The intriguing magnetic properties of nanoscale semiconductors have profound technological implications in various fields. These implications include information storage and processing, spintronics, magnetic sensors, and biomedical applications.

Nanomagnetic materials can be utilized in the development of high-density magnetic storage devices, such as hard drives, where information is stored and retrieved based on the orientation of nanoscale magnetic domains. Spintronics, or spin-based electronics, exploit the intrinsic spin of electrons to enable new functionalities in electronic devices, such as non-volatile memory and spin logic gates.

Furthermore, nanomagnetism plays a crucial role in the development of magnetic sensors with high sensitivity and spatial resolution. These sensors find applications in diverse areas, such as navigation, medical imaging, and non-destructive testing.

In the biomedical field, nanomagnetic materials offer prospects for targeted drug delivery, magnetic hyperthermia cancer therapy, and magnetic resonance imaging contrast agents, leveraging their unique magnetic properties at the nanoscale.

Conclusion

In this chapter, we have explored the captivating field of nanomagnetism and its relevance in understanding magnetic behaviors in nanoscale semiconductors. Through an understanding of magnetic moments, spins, exchange interactions, magnetic anisotropy, and the dynamics of nanomagnetism, we gain valuable insights into the technological implications of these magnetic properties. Nanomagnetic materials open up exciting avenues in various fields, ranging from information storage to biomedical applications. The mathematical descriptions and theoretical frameworks presented in this chapter lay the foundation for further research and exploration in the ever-expanding field of nanomagnetism.

Python Code Snippet

Below is a Python code snippet that calculates exchange interaction energy, magnetic anisotropy energy, and the dynamics of magnetization in nanomagnetism.

```python
import numpy as np

def calculate_exchange_energy(J, S):
    '''
    Calculate the exchange interaction energy.
    :param J: Exchange integral (Joules).
    :param S: List of spin vectors for interacting spins.
    :return: Exchange interaction energy (Joules).
    '''
```

```python
    energy = -J * np.sum(np.dot(S[i], S[j]) for i in range(len(S))
              for j in range(i + 1, len(S)))
    return energy

def calculate_anisotropy_energy(K, theta):
    '''
    Calculate the magnetic anisotropy energy.
    :param K: Anisotropy constant (Joules per cubic meter).
    :param theta: Angle between magnetization vector and preferred
              direction (radians).
    :return: Anisotropy energy (Joules).
    '''
    anisotropy_energy = K * np.sin(theta)**2
    return anisotropy_energy

def landau_lifshitz_gilbert(M, H_eff, gamma, alpha, dt):
    '''
    Update the magnetization vector using the
              Landau-Lifshitz-Gilbert equation.
    :param M: Current magnetization vector (numpy array).
    :param H_eff: Effective magnetic field vector (numpy array).
    :param gamma: Gyromagnetic ratio (radians per second per Tesla).
    :param alpha: Damping parameter (dimensionless).
    :param dt: Time step for the update (seconds).
    :return: Updated magnetization vector (numpy array).
    '''
    M = M / np.linalg.norm(M)  # Normalize M
    dM_dt = -gamma * np.cross(M, H_eff) + alpha * np.cross(M,
              np.cross(M, H_eff))
    M_new = M + dM_dt * dt  # Update magnetization
    return M_new / np.linalg.norm(M_new)  # Normalize again

# Constants and inputs for the calculations
J = 1e-20  # Example exchange integral in Joules
S = [np.array([1, 0, 0]), np.array([-1, 0, 0])]  # Two spin vectors
K = 1e-21  # Anisotropy constant in Joules per cubic meter
theta = np.pi / 4  # 45 degrees in radians
gamma = 2.21e5  # Gyromagnetic ratio for typical ferromagnets in
              rad/(s·T)
alpha = 0.1  # Damping parameter
dt = 1e-12  # Time step in seconds
M_initial = np.array([0, 0, 1])  # Initial magnetization vector in
              the z-direction
H_eff = np.array([0, 0, 1])  # Effective magnetic field vector

# Calculations
exchange_energy = calculate_exchange_energy(J, S)
anisotropy_energy = calculate_anisotropy_energy(K, theta)
M_updated = landau_lifshitz_gilbert(M_initial, H_eff, gamma, alpha,
              dt)

# Output results
print("Exchange Interaction Energy:", exchange_energy, "Joules")
```

```
print("Magnetic Anisotropy Energy:", anisotropy_energy, "Joules")
print("Updated Magnetization Vector:", M_updated)
```

This code defines three functions:

- `calculate_exchange_energy` computes the exchange interaction energy based on the exchange integral and the spin vectors.
- `calculate_anisotropy_energy` calculates the magnetic anisotropy energy given the anisotropy constant and the angle.
- `landau_lifshitz_gilbert` updates the magnetization vector using the LLG equation based on the current magnetization, effective magnetic field, gyromagnetic ratio, damping parameter, and time step.

The provided example calculates the exchange interaction energy, magnetic anisotropy energy, and updates the magnetization vector, then prints the results.

Chapter 26

Spin Dynamics in Nanostructures

Spin dynamics plays a crucial role in understanding and modeling the behavior of nanostructured materials, specifically their spin-related properties and effects. In this chapter, we delve into the mathematical description and theoretical frameworks that enable us to investigate the role of spin and its dynamics in nanostructured materials. By developing a comprehensive understanding of spin dynamics, we can gain valuable insights into the behavior and potential applications of these materials.

Spin and Its Quantum Description

In the realm of quantum mechanics, the spin of a particle is a fundamental quantum property that gives rise to a magnetic moment. It is characterized by a spin quantum number, denoted by s, which determines the allowed values of the spin projection. To describe the quantum state of a spin-s particle, we use a mathematical formalism known as spinors.

A spinor is a mathematical object that represents the wavefunction of a spin-s particle. It is usually denoted by a column matrix with $2s+1$ complex components. For example, a spin-$\frac{1}{2}$ particle, such as an electron, can be described by a two-component spinor.

The time evolution of a spin-s particle is governed by the Schrödinger equation, which can be written as:

$$i\hbar \frac{\partial}{\partial t}\psi(t) = \hat{H}\psi(t)$$

where $\hat{H}$ is the Hamiltonian operator and $\psi(t)$ is the spinor representing the quantum state of the particle at time t.

Spin Dynamics in Magnetic Fields

When nanostructured materials are subjected to external magnetic fields, the dynamics of their spins are significantly influenced. The interaction between the spin of the particles and the magnetic field can give rise to precession, damping, and other spin-related phenomena.

The precession of spin can be described by the Bloch equations, which govern the time evolution of the spin density operator. In the presence of an external magnetic field **B**, the Bloch equations can be written as a set of coupled differential equations:

$$\frac{dM_x}{dt} = \gamma(M_y B_z - M_z B_y) - R_x M_x$$

$$\frac{dM_y}{dt} = \gamma(M_z B_x - M_x B_z) - R_y M_y$$

$$\frac{dM_z}{dt} = \gamma(M_x B_y - M_y B_x) - R_z(M_z - M_{z0})$$

where M_x, M_y, and M_z are the components of the magnetization vector, γ is the gyromagnetic ratio, **B** is the external magnetic field vector, and R_x, R_y, and R_z are relaxation rates. The term $(M_z - M_{z0})$ represents the deviation of the z-component of the magnetization from its equilibrium value M_{z0}.

These equations describe the precession of spins around the effective magnetic field $\mathbf{B}_{\text{eff}}$, as well as the relaxation of the magnetization towards its equilibrium value.

Spin-Orbit Interaction

The spin dynamics in nanostructured materials can also be significantly influenced by spin-orbit interactions. These interactions arise from the coupling between the spin of the particles and their orbital motion. The spin-orbit interaction can lead to spin-flip processes, spin relaxation, and other spin-related phenomena.

Mathematically, the spin dynamics in the presence of spin-orbit interactions can be described using the Rashba Hamiltonian. The Rashba Hamiltonian is given by:

$$\hat{H}_{\text{Rashba}} = \frac{\alpha}{\hbar}(\mathbf{p} \times \sigma) \cdot \mathbf{z}$$

where α is the Rashba coupling strength, $\mathbf{p}$ is the momentum operator, σ is the vector of Pauli spin matrices, and $\mathbf{z}$ is the unit vector perpendicular to the plane of motion.

The Rashba Hamiltonian describes the coupling between the spin of the particles and their momentum in the presence of an electric field gradient. It can give rise to spin-dependent energy shifts and induce spin precession.

Spin Transport and Spin Currents

Understanding spin transport and spin currents in nanostructured materials is of great importance for spintronics and other emerging technologies. Spin currents represent the flow of spins in a material, and their dynamics can be influenced by various factors, such as spin-orbit interactions, magnetic fields, and scattering processes.

The spin dynamics in the presence of spin currents can be described using the spin continuity equation. The spin continuity equation is given by:

$$\frac{\partial \mathbf{S}}{\partial t} + \nabla \cdot \mathbf{j}_s = \frac{\mathbf{S}}{\tau_s}$$

where $\mathbf{S}$ is the spin density vector, $\mathbf{j}_s$ is the spin current density vector, and τ_s is the spin relaxation time.

This equation describes the conservation of spin in a material, accounting for the generation, propagation, and relaxation of spin currents. Solving the spin continuity equation allows us to analyze the dynamics of spin transport and investigate phenomena such as spin accumulation and spin Hall effect.

Conclusion

In this chapter, we have explored the intriguing world of spin dynamics in nanostructured materials. By understanding the mathematical description and theoretical frameworks of spin dynamics, we gain valuable insights into the behavior of spins and their dynamics in various nanostructures. The mathematical formalism of spinors, coupled with the Bloch equations, allow us to describe the precession and relaxation of spins in magnetic fields. Spin-orbit interactions and their impact on spin dynamics were also discussed, along with the Rashba Hamiltonian. Lastly, the spin transport and spin continuity equation provided a framework for investigating spin currents in nanostructures. By advancing our understanding of spin dynamics, we can pave the way for the development of novel spin-based technologies and devices.

Python Code Snippet

Below is a Python code snippet that calculates the dynamics of spins in nanostructured materials, including precession in a magnetic field and spin transport through the spin continuity equation.

```python
import numpy as np
import matplotlib.pyplot as plt

def bloch_equations(M, B, gamma, R, dt):
    '''
    Update the magnetization vector M based on the Bloch equations.
    :param M: Current magnetization vector [M_x, M_y, M_z].
    :param B: External magnetic field vector [B_x, B_y, B_z].
    :param gamma: Gyromagnetic ratio.
    :param R: Relaxation rates [R_x, R_y, R_z].
    :param dt: Time step for the simulation.
    :return: Updated magnetization vector.
    '''
    M_x, M_y, M_z = M
    B_x, B_y, B_z = B
    R_x, R_y, R_z = R

    dM_x = gamma * (M_y * B_z - M_z * B_y) - R_x * M_x
    dM_y = gamma * (M_z * B_x - M_x * B_z) - R_y * M_y
    dM_z = gamma * (M_x * B_y - M_y * B_x) - R_z * (M_z - 1)  #
    ↪  M_{z0} is taken as 1 for normalized case

    return np.array([M_x + dM_x * dt, M_y + dM_y * dt, M_z + dM_z *
    ↪  dt])

def simulate_spin_dynamics(B, gamma, R, M0, time_span, dt):
    '''
    Simulate spin dynamics over a given time span.
    :param B: External magnetic field vector [B_x, B_y, B_z].
    :param gamma: Gyromagnetic ratio.
```

```
:param R: Relaxation rates [R_x, R_y, R_z].
:param M0: Initial magnetization vector [M_x0, M_y0, M_z0].
:param time_span: Total time of simulation.
:param dt: Time step for the simulation.
:return: Arrays of time and magnetization over the simulation.
'''
time_array = np.arange(0, time_span, dt)
M = M0
M_list = []

for t in time_array:
    M = bloch_equations(M, B, gamma, R, dt)
    M_list.append(M)

return time_array, np.array(M_list)

# Define constants
gamma = 1.76e11   # Gyromagnetic ratio for protons in rad/s/T
B = np.array([0, 0, 1])   # External magnetic field in Tesla
R = np.array([0.1, 0.1, 0.1])   # Relaxation rates
M0 = np.array([0, 0, 1])   # Initial magnetization vector
time_span = 0.1   # Total time of simulation in seconds
dt = 0.001   # Time step in seconds

# Simulate spin dynamics
time_array, M_list = simulate_spin_dynamics(B, gamma, R, M0,
     time_span, dt)

# Plot the results
plt.figure(figsize=(10, 6))
plt.plot(time_array, M_list[:, 0], label='M_x', color='blue')
plt.plot(time_array, M_list[:, 1], label='M_y', color='orange')
plt.plot(time_array, M_list[:, 2], label='M_z', color='green')
plt.xlabel('Time (s)')
plt.ylabel('Magnetization')
plt.title('Spin Dynamics in a Magnetic Field')
plt.legend()
plt.grid()
plt.show()
```

This code defines two functions:

- `bloch_equations` calculates the update of the magnetization vector based on the Bloch equations under the influence of an external magnetic field and relaxation processes.
- `simulate_spin_dynamics` runs a simulation of spin dynamics over a specified time span, returning the time and magnetization values.

The provided example simulates the dynamics of spins in a magnetic field, demonstrating how magnetization components change over time due to external influences and relaxation effects, then visualizes the results using a plot.

Chapter 27

Charge Trapping and Defects

In this chapter, we delve into the realm of charge trapping mechanisms and defects in nanoscale semiconductors. With a focus on detailed modeling, we aim to provide expert insight into the nature and impact of these phenomena in the context of nanoscale semiconductor materials.

Introduction

Charge trapping and defects are ubiquitous in nanoscale semiconductors and play a crucial role in their electrical and optical properties. Understanding and modeling these phenomena is of paramount importance for the design and optimization of semiconductor devices. In this section, we provide an introduction to charge trapping mechanisms and defects, laying the foundation for the subsequent sections.

Charge Trapping Mechanisms

Charge trapping refers to the process by which charge carriers, such as electrons and holes, become immobilized in localized states within the semiconductor material. These trapped charges can significantly impact the performance of semiconductor devices, introducing effects such as carrier lifetime reduction, increased recombination rates, and altered electronic transport properties. In this section, we explore the various mechanisms responsible for charge trapping in nanoscale semiconductors.

1 Trap States and Energy Levels

At the heart of charge trapping phenomena are trap states within the energy bandgap of the semiconductor material. These trap states can arise from a variety of sources, including impurities, crystal defects, and surface/interface states. Each trap state possesses an associated energy level, which determines the probability of charge carriers occupying the state. The distribution of trap states and their energy levels play a crucial role in charge trapping dynamics and device performance.

2 Carrier Capture and Emission Processes

Charge carriers can become trapped by undergoing capture processes, where they transition from the conducting band to trap states. Similarly, trapped carriers can be released back to the conducting band through emission processes. These capture and emission processes are influenced by factors such as carrier energy, trap density of states, and thermal energy. In this subsection, we delve into the mathematical description of carrier capture and emission processes, offering a detailed modeling framework.

3 Trap Filling and Emptying Dynamics

The occupancy of trap states is determined by a delicate balance between carrier capture and emission processes. Trap filling occurs when the rate of carrier capture exceeds that of emission, leading to an increase in trapped charge density. Conversely, trap emptying occurs when the rate of emission exceeds that of capture, resulting in a decrease in trapped charge density. Understanding the dynamics of trap filling and emptying is crucial for predicting and optimizing the behavior of trapped charge carriers.

Defects in Nanoscale Semiconductors

In addition to charge trapping, defects in the crystal structure of nanoscale semiconductors can profoundly influence their electrical and optical properties. Defects arise from imperfections such as missing atoms, impurities, and lattice distortions. In this section, we explore the nature and impact of defects in nanoscale semiconductors.

1 Defect Energy Levels

Defects introduce energy levels within the energy bandgap of the semiconductor, modifying the electronic structure and introducing localized states. These defect energy levels can act as trapping centers for charge carriers, leading to charge trapping phenomena discussed earlier. The distribution and nature of defect energy levels govern the behavior of trapped charges and carrier dynamics.

2 Defect-Induced Strain and Carrier Mobility

Defects can also introduce strain into the crystal lattice of semiconductors, altering the band structure and carrier mobility. Strained regions surrounding defects can result in localized electronic states and scattering of charge carriers, thereby influencing carrier transport properties. Modeling the impact of defect-induced strain requires a combination of quantum mechanical calculations and continuum mechanics.

3 Impact on Device Performance

The presence of charge trapping and defects in nanoscale semiconductors has profound implications for the performance of semiconductor devices. These phenomena can degrade carrier mobility, increase leakage current, reduce device lifetime, and alter the emission properties of optoelectronic devices. Understanding the impact of charge trapping and defects on device performance is essential for device optimization and reliability.

Conclusion

In this chapter, we have provided detailed modeling approaches for understanding charge trapping mechanisms and defects in nanoscale semiconductors. By exploring the physics behind charge trapping phenomena and the impact of defects, we have gained valuable insight into their influence on the electrical and optical properties of nanoscale semiconductors. Through a combination of theoretical modeling and experimental characterization, we can further advance our understanding of charge trapping and defects, paving the way for improved nanoscale semiconductor devices.

Python Code Snippet

Below is a Python code snippet that models charge trapping mechanisms and defect dynamics in nanoscale semiconductors, implementing key equations and algorithms presented in this chapter.

```python
import numpy as np

def calculate_trap_fill_rate(trap_density, capture_cross_section,
     carrier_flux):
    '''
    Calculate the rate of trap filling based on trap density and
      carrier dynamics.
    :param trap_density: Density of trap states in m^-3.
    :param capture_cross_section: Capture cross-section area in m^2.
    :param carrier_flux: Flux of carriers in m^-2 s^-1.
    :return: Trap fill rate in m^-3 s^-1.
    '''
    return trap_density * capture_cross_section * carrier_flux


def calculate_trap_occupancy(capture_rate, emission_rate,
     transient_time):
    '''
    Calculate the occupancy of trap states over a certain transient
      time.
    :param capture_rate: Rate at which carriers are captured in m^-3
      s^-1.
    :param emission_rate: Rate at which carriers are emitted in m^-3
      s^-1.
    :param transient_time: Time duration to observe occupancy in
      seconds.
    :return: Occupancy of the trap states in m^-3.
    '''
    occupancy = (capture_rate - emission_rate) * transient_time
    return max(occupancy, 0)  # Occupancy cannot be negative


def defect_energy_level(defect_type, energy_gap):
    '''
    Determine the energy level introduced by a defect type within
      the energy gap.
    :param defect_type: Type of defect (e.g., "donor", "acceptor").
    :param energy_gap: Energy gap of the semiconductor in eV.
    :return: Defect energy level in eV.
    '''
```

```python
    if defect_type == "donor":
        return energy_gap / 2 + 0.1  # Example: donor level slightly
        ↪ above midgap
    elif defect_type == "acceptor":
        return energy_gap / 2 - 0.1  # Example: acceptor level
        ↪ slightly below midgap
    else:
        raise ValueError("Invalid defect type. Use 'donor' or
        ↪ 'acceptor'.")


# Constants and parameters for simulation
trap_density = 1e20    # Trap state density in m^-3
capture_cross_section = 1e-20    # Capture cross-section in m^2
carrier_flux = 1e25    # Carrier flux in m^-2 s^-1
transient_time = 1e-6    # Time in seconds for occupancy calculation
emission_rate = 1e18   # Emission rate in m^-3 s^-1
defect_type = "donor"    # Type of defect in the material
energy_gap = 1.1    # Energy gap of the semiconductor in eV

# Calculations
trap_fill_rate = calculate_trap_fill_rate(trap_density,
↪    capture_cross_section, carrier_flux)
trap_occupancy = calculate_trap_occupancy(trap_fill_rate,
↪    emission_rate, transient_time)
defect_level = defect_energy_level(defect_type, energy_gap)

# Output results
print("Trap Fill Rate:", trap_fill_rate, "m^-3 s^-1")
print("Trap Occupancy:", trap_occupancy, "m^-3")
print("Defect Energy Level for type", defect_type + ":",
↪    defect_level, "eV")
```

This code defines three functions:

- `calculate_trap_fill_rate` computes the trap fill rate based on the density of trap states, capture cross-section, and carrier flux.
- `calculate_trap_occupancy` calculates the occupancy of the trap states over a specified transient time by evaluating capture and emission rates.
- `defect_energy_level` determines the defect energy level within the bandgap depending on the type of defect (donor or acceptor).

The provided example performs calculations for trap fill rate, trap occupancy, and defect energy levels, then prints the results.

Chapter 28

Nanostructured Field-Effect Devices

Introduction

In this chapter, we delve into the realm of nanostructured field-effect devices and explore the benefits that nanostructuring brings to this technology. Field-effect devices, such as field-effect transistors (FETs), are fundamental building blocks of modern electronic circuits. By miniaturizing and manipulating the structures at the nanoscale, we can enhance their performance and unlock new functionalities. In this section, we provide an overview of the field-effect devices, highlighting their importance in the field of nanoelectronics.

Fundamentals of Field-Effect Devices

Field-effect devices function based on the modulation of a charge carrier density in a semiconductor channel by an external electric field. This modulation is achieved through the control of a gate electrode, which creates an electric field that allows or blocks the flow of charge carriers in the channel. The most common type of field-effect device is the field-effect transistor (FET), which serves as a crucial component in digital circuits, analog amplifiers, and many other electronic applications.

1 Basic Operation Principles

The basic operation of an FET involves the manipulation of charge carriers through the application of a gate voltage. By adjusting this voltage, the device can be switched between an "on" state, where a conductive channel is formed, and an "off" state, where the channel is effectively turned off. This control over charge carrier flow enables the amplification and switching of electrical signals. The key parameters characterizing the performance of an FET include the on-off current ratio, the threshold voltage, and the subthreshold slope.

2 Traditional FET Structures

Traditional FETs utilize a planar structure, with a gate electrode deposited on top of a planar semiconductor channel. The channel material determines the device's electrical properties, such as carrier mobility and threshold voltage. Silicon-based FETs, including metal-oxide-semiconductor FETs (MOSFETs), have dominated the electronics industry for decades due to the excellent properties of silicon and the well-established fabrication technologies. However, as device dimensions approach the nanoscale, new challenges arise, such as short-channel effects, leakage currents, and limited scalability. Nanostructuring techniques provide solutions to overcome these challenges and further enhance FET performance.

Benefits of Nanostructuring in FET Technology

Nanostructuring enables the fabrication of field-effect devices with enhanced properties and novel functionalities. By manipulating the device's geometrical and physical characteristics at the nanoscale, we can overcome the limitations of traditional FETs and unlock new opportunities. In this section, we explore the key benefits that nanostructuring brings to FET technology.

1 Improved Control Over Channel Conductance

Nanoscale FETs offer enhanced control over the conductance of the semiconductor channel. By reducing the channel length, the gate can exert a stronger influence on the charge carriers, resulting in improved on-off current ratios and faster switching speeds. Additionally, the introduction of nanowires or nanosheets as the channel material can boost carrier mobility, leading to higher device performance.

2 Enhanced Electrostatic Control

Nanostructured FETs also provide improved electrostatic control over the channel region. By scaling down the device dimensions, the gate electrode can exert a more significant electrostatic force on the charge carriers, leading to enhanced gate control efficiency. This enhanced control allows for the reduction of the threshold voltage and the suppression of short-channel effects, thereby improving device performance and reducing power consumption.

3 Novel Device Architectures

Nanostructuring techniques enable the realization of novel device architectures that go beyond the limitations of traditional FETs. Examples include gate-all-around (GAA) FETs, nanowire FETs, and tunnel FETs. These alternative architectures offer unique advantages, such as improved electrostatic control, reduced leakage current, and enhanced device performance. By exploring these novel device structures, researchers can push the boundaries of FET technology and develop innovative applications.

4 Integration with Other Nanomaterials

Nanostructured FETs can be seamlessly integrated with other nanomaterials and nanostructures, such as nanowires, nanotubes, and two-dimensional materials. This integration enables the exploitation of synergistic effects and the realization of hybrid systems with enhanced functionalities. For example, the combination of a nanostructured FET with a plasmonic material can enable enhanced light-matter interactions and novel sensing capabilities.

Conclusion

Nanostructured field-effect devices offer significant advantages over traditional FETs, including improved control over channel conductance, enhanced electrostatic control, novel device architectures, and seamless integration with other nanomaterials. These benefits pave the way for the development of high-performance electronic devices and open up new opportunities in the field of nanoelectronics. Through the utilization of nanostructuring techniques, researchers and engineers can continue to innovate and push the boundaries of FET technology, leading to advancements in various electronic applications.

Python Code Snippet

Below is a Python code snippet that models the operation of a nanostructured field-effect transistor (FET), including calculations for the on-off current ratio, threshold voltage, and subthreshold slope.

```
import numpy as np
import matplotlib.pyplot as plt

def calculate_current(v_g, v_d, v_th, carrier_mobility):
    '''
    Calculate the drain current (I_d) in a nanoscale FET based on
    ↪ gate voltage (v_g),
    drain voltage (v_d), threshold voltage (v_th), and carrier
    ↪ mobility.

    :param v_g: Gate voltage in volts.
    :param v_d: Drain voltage in volts.
    :param v_th: Threshold voltage in volts.
    :param carrier_mobility: Carrier mobility in m^2/(V·s).
    :return: Drain current in amperes.
    '''
    if v_g < v_th:
        return 0  # Device is OFF if gate voltage is below threshold
    # Saturation region current calculation
    I_d = (carrier_mobility / 2) * (v_g - v_th)**2
    return I_d

def calculate_on_off_current_ratio(I_on, I_off):
    '''
    Calculate the on-off current ratio of a FET.
    :param I_on: Drain current in the "on" state in amperes.
    :param I_off: Drain current in the "off" state in amperes.
    :return: On-off current ratio (dimensionless).
    '''
    return I_on / I_off if I_off != 0 else float('inf')

def calculate_subthreshold_slope(I_d, v_g, v_th):
    '''
    Calculate subthreshold slope of the device.
```

```
    :param I_d: Drain current in amperes.
    :param v_g: Gate voltage in volts.
    :param v_th: Threshold voltage in volts.
    :return: Subthreshold slope in mV/decade.
    '''
    return 60 / np.log10(I_d) * (v_g - v_th)

# Parameters
v_th = 0.5  # Threshold voltage in volts
carrier_mobility = 200e-4  # Carrier mobility in m^2/(V·s)
v_g_values = np.linspace(0, 1, 100)  # Gate voltage values from 0V
↪ to 1V
v_d = 0.5  # Drain voltage in volts

# Calculate currents for each gate voltage
I_d_values = [calculate_current(v_g, v_d, v_th, carrier_mobility)
↪ for v_g in v_g_values]
I_on = I_d_values[-1]  # Drain current in the "on" state
I_off = I_d_values[0]  # Drain current in the "off" state

# Calculate on-off current ratio
on_off_ratio = calculate_on_off_current_ratio(I_on, I_off)

# Calculate subthreshold slope at threshold voltage
subthreshold_slope = calculate_subthreshold_slope(I_off, v_th, v_th)

# Output results
print("Drain current in 'on' state:", I_on, "A")
print("Drain current in 'off' state:", I_off, "A")
print("On-Off Current Ratio:", on_off_ratio)
print("Subthreshold Slope:", subthreshold_slope, "mV/decade")

# Plotting the characteristics
plt.figure(figsize=(10, 5))
plt.plot(v_g_values, I_d_values, label='I_d vs. V_g', color='blue')
plt.axhline(y=I_th, color='r', linestyle='--', label='Threshold
↪ current (I_th)')
plt.axvline(x=v_th, color='green', linestyle='--', label='Threshold
↪ voltage (V_th)')
plt.title('Drain Current Characteristics of Nanostructured FET')
plt.xlabel('Gate Voltage (V)')
plt.ylabel('Drain Current (A)')
plt.legend()
plt.grid()
plt.yscale('log')
plt.show()
```

This code defines three functions:

- **calculate_current** computes the drain current based on gate voltage, drain voltage, threshold voltage, and carrier mobility for

a nanoscale FET.
- `calculate_on_off_current_ratio` determines the on-off current ratio of the device.
- `calculate_subthreshold_slope` evaluates the subthreshold slope for the device based on drain current and gate voltage.

The example provided calculates the drain current in both the "on" and "off" states, the on-off current ratio, and the subthreshold slope, then visualizes the characteristics of the FET with a plot.

Chapter 29

Multi-Scale Modeling Approaches

Introduction

In the field of semiconductor modeling, accurately capturing the behavior of materials and devices across multiple length scales is crucial for understanding their properties and predicting their performance. Multi-scale modeling approaches aim to integrate various scales, ranging from the atomic level to the macroscopic level, in order to simulate semiconductor systems comprehensively. This chapter explores the challenges and methodologies involved in bridging the gap between different scales of semiconductor modeling, with a focus on the integration of atomic, mesoscale, and macroscopic models.

Atomic Scale Modeling

At the atomic scale, semiconductor materials are represented by their crystal structures and the arrangement of atoms within them. In this regime, density functional theory (DFT) and molecular dynamics (MD) simulations are commonly used to calculate the electronic structure, energetics, and dynamics of semiconductor systems. These methods rely on solving the Schrödinger equation for electrons or simulating the motion of atoms according to classical equations of motion. The resulting information, such as band structures and atomic trajectories, provides insights into the microscopic behavior of semiconductors.

Mesoscale Modeling

The mesoscale bridges the gap between the atomic and macroscopic scales, capturing the interactions and phenomena that emerge at intermediate length scales. In semiconductor modeling, mesoscale approaches are employed to study phenomena such as charge transport, defect formation, and material growth. Continuum-based models, such as the drift-diffusion equation and the Poisson equation, are widely used to describe carrier transport and electrostatics at this scale. Additionally, kinetic Monte Carlo methods and phase-field models offer valuable means to simulate the dynamics of defects and interface evolution in semiconductor devices.

Macroscopic Scale Modeling

At the macroscopic scale, semiconductor modeling focuses on capturing the overall behavior of devices and systems. Here, continuum mechanics and Maxwell's equations are employed to describe the transport of charge carriers, heat transfer, and electromagnetic phenomena. This level of modeling often involves solving partial differential equations (PDEs) to simulate the behavior of semiconductor devices, such as field-effect transistors (FETs) and solar cells. Finite element methods (FEM) and finite difference methods (FDM) are commonly used numerical techniques for solving these PDEs and obtaining meaningful macroscopic insights.

Integration of Multi-Scale Models

Integrating models across multiple scales is a challenging task that requires careful consideration of the interfaces and interactions between the different scales. Several approaches have been developed to achieve this integration effectively. One common strategy is the hierarchical modeling approach, where different models are employed at each scale, and information is exchanged across scales through coupling conditions. Another approach involves the use of hybrid models, combining atomistic simulations with continuum models to capture both local atomic details and macroscopic behavior simultaneously. Additionally, model reduction techniques, such as coarse-graining and homogenization, can be employed to simplify the complexity of multi-scale models and make them computationally tractable.

Conclusion

Multi-scale modeling approaches enable a comprehensive understanding of semiconductor systems by integrating models across atomic, mesoscale, and macroscopic scales. From the atomic scale, where DFT and MD simulations provide insights into electronic and atomic behavior, to the mesoscale, where continuum-based models capture transport phenomena and defect dynamics, to the macroscopic scale, where PDEs and numerical methods describe overall device behavior, these approaches allow for the exploration and prediction of semiconductor properties and device performance. By bridging the gap between scales, multi-scale modeling paves the way for advancements in semiconductor technology and the design of innovative devices.

Python Code Snippet

Below is a Python code snippet that implements important equations and algorithms related to multi-scale modeling in semiconductor physics, including atomic, mesoscale, and macroscopic scale models.

```python
import numpy as np
from scipy.integrate import solve_ivp
import matplotlib.pyplot as plt

# Constants
electron_charge = 1.6e-19  # Charge of an electron in Coulombs
epsilon_0 = 8.85419e-12  # Permittivity of free space in F/m

def calculate_band_structure(energies, k_points):
    '''
    Calculate the density of states (DOS) for a given band
    structure.
    :param energies: List of energy levels.
    :param k_points: Number of k-points in the Brillouin zone.
    :return: Density of states.
    '''
```

```python
    dos = np.histogram(energies, bins=100, density=True)[0]
    return dos

def drift_diffusion_equations(t, y, D, mu, n0):
    '''
    A system of ordinary differential equations for the
    ↪ drift-diffusion model.
    :param t: Time variable.
    :param y: Array of carrier concentrations and electric field.
    :param D: Diffusion coefficient.
    :param mu: Mobility.
    :param n0: Initial carrier concentration.
    :return: Derivative of carrier concentrations and electric
    ↪ field.
    '''
    n, E = y
    dn_dt = D * (n0 - n) - mu * E * n  # Drift-Diffusion equation
    ↪ for electron concentration
    dE_dt = -n  # Poisson's equation for electric field
    return [dn_dt, dE_dt]

def simulate_drift_diffusion(D, mu, n0, t_span):
    '''
    Simulate the drift-diffusion behavior over time.
    :param D: Diffusion coefficient.
    :param mu: Mobility.
    :param n0: Initial carrier concentration.
    :param t_span: Time span for the simulation.
    :return: Time array and concentration/field array.
    '''
    sol = solve_ivp(drift_diffusion_equations, t_span, [n0, 0],
    ↪  args=(D, mu, n0), t_eval=np.linspace(t_span[0], t_span[1],
    ↪  100))
    return sol.t, sol.y

def plot_results(t, n, E):
    '''
    Plot the results of the drift-diffusion simulation.
    :param t: Time array.
    :param n: Electron concentration over time.
    :param E: Electric field over time.
    '''
    plt.figure(figsize=(10, 5))
    plt.subplot(1, 2, 1)
    plt.plot(t, n, label='Electron Concentration')
    plt.title('Drift-Diffusion Simulation - Electron Concentration')
    plt.xlabel('Time (s)')
    plt.ylabel('Concentration (m^-3)')
    plt.grid(True)

    plt.subplot(1, 2, 2)
    plt.plot(t, E, label='Electric Field', color='orange')
    plt.title('Drift-Diffusion Simulation - Electric Field')
```

```
    plt.xlabel('Time (s)')
    plt.ylabel('Electric Field (V/m)')
    plt.grid(True)

    plt.tight_layout()
    plt.show()

# Parameters for simulation
D = 1e-5     # Diffusion coefficient in m^2/s
mu = 1e-2    # Mobility in m^2/(V·s)
n0 = 1e20    # Initial carrier concentration in m^-3
t_span = (0, 1e-3)  # Time span for simulation in seconds

# Perform simulation
t, y = simulate_drift_diffusion(D, mu, n0, t_span)

# Plot results
plot_results(t, y[0], y[1])
```

This code defines several methods for simulating key phenomena in semiconductor physics:

- `calculate_band_structure` computes the density of states for given energy levels.
- `drift_diffusion_equations` provides a set of ODEs for the drift-diffusion model.
- `simulate_drift_diffusion` runs a simulation of the drift-diffusion behavior using 'scipy''s integrator.
- `plot_results` visualizes the results of the simulation, displaying both electron concentration and electric field over time.

In this example, a drift-diffusion model for electron behavior in a semiconductor is calculated and plotted, providing insights into the temporal evolution of carrier concentration and electric field.

Chapter 30

Nanostructured LEDs and Lasers

The field of nanostructured LEDs and lasers has gained significant attention in recent years due to the promising potential for enhancing light emission through structural nano-engineering. In this chapter, we delve into the simulation techniques used to model and analyze the behavior of nanostructured devices. Specifically, we focus on the simulation of light emission enhancements in nanostructured LEDs and lasers.

Introduction

Nanostructured LEDs and lasers have emerged as powerful tools for controlling and manipulating the emission of light at the nanoscale. By carefully designing the structures and properties of semiconductor materials, it is possible to achieve enhanced light emission efficiency and control over the emitted light's properties, such as intensity, wavelength, and polarization. Understanding and predicting the behavior of these devices requires sophisticated simulation techniques that can capture the complex physical phenomena occurring at the nanoscale.

Electro-optical Modeling of Nanostructured LEDs

The electro-optical behavior of nanostructured LEDs can be effectively modeled using advanced simulation techniques. One commonly employed approach is the rate equation model, which describes the carrier dynamics and light emission in the device. The rate equation model is based on solving a set of coupled differential equations to calculate the carrier concentrations and the photon density. The equations governing the carrier dynamics can be written as follows:

$$\frac{dN_p}{dt} = G - \frac{N_p}{\tau_p} - \frac{N_p}{\tau_s} \quad (30.1)$$

$$\frac{dN_n}{dt} = G - \frac{N_n}{\tau_n} - \frac{N_n}{\tau_s} \quad (30.2)$$

where N_p and N_n represent the concentrations of holes and electrons, respectively. τ_p, τ_n, and τ_s denote the carrier lifetimes for holes, electrons, and photons, respectively. G represents the generation rate of electron-hole pairs due to electrical or optical excitation.

To calculate the photon density, an optical model is required. One widely used model is the rate equation model for photon density, which can be expressed as:

$$\frac{dS}{dt} = \frac{N_p}{\tau_p} - \frac{S}{\tau_s} + R_{sp} - R_{sp}^{nonrad} \quad (30.3)$$

where S represents the photon density, R_{sp} is the radiative recombination rate, and R_{sp}^{nonrad} is the nonradiative recombination rate.

These coupled rate equations provide a comprehensive framework for simulating the electro-optical behavior of nanostructured LEDs, allowing researchers to study the effects of various structural modifications on light emission efficiency.

Laser Modeling in Nanostructured Devices

The modeling of lasers in nanostructured devices requires a more sophisticated approach due to the intricate interplay between optical and electrical phenomena. A comprehensive laser model typically consists of coupled equations that describe the gain and loss mechanisms in the device.

One key component of laser modeling is the calculation of the optical gain. The optical gain is determined by the carrier distribution and the material properties of the active medium. A common approach is to use the semiconductor Bloch equations, which describe the carrier dynamics in an active medium subjected to an applied electric field. These equations can be written as:

$$\frac{dn_k}{dt} = \frac{i}{\hbar}(E_k - i\Gamma_k)n_k + \left(\frac{dn_k}{dt}\right)_{coll} \tag{30.4}$$

where n_k represents the electron occupation probability of the kth energy level, E_k is the energy level, Γ_k is the dephasing rate, and $\left(\frac{dn_k}{dt}\right)_{coll}$ denotes the carrier relaxation due to carrier-carrier and carrier-phonon collisions.

To understand the laser's behavior, the coupled rate and propagation equations are needed. The rate equations are similar to the ones used in LED modeling, accounting for carrier dynamics, radiative recombination, and gain. The propagation equation, commonly known as the wave equation, describes the propagation of the optical field inside the laser cavity and can be expressed as:

$$\frac{\partial E}{\partial t} = (\delta\nu + g)E + \frac{i\alpha}{2}E + \sqrt{\frac{2\alpha P_{in}}{n_g A}} + \sqrt{2\gamma}\tilde{s} \tag{30.5}$$

where E is the electric field envelope, $\delta\nu$ is the frequency detuning, g represents the gain, α is the absorption coefficient, P_{in} denotes the input power, n_g is the group refractive index, A represents the mode area, γ is the spontaneous emission factor, and $\tilde{s}$ represents the quantum noise.

The coupled laser model provides a rigorous framework for simulating the behavior of nanostructured lasers, allowing researchers to optimize device designs for improved performance and efficiency.

Simulation Techniques

To solve the complex systems of equations involved in the modeling of nanostructured LEDs and lasers, sophisticated simulation techniques are employed. Numerical methods such as finite difference methods (FDM), finite element methods (FEM), and finite volume methods (FVM) play a crucial role in solving these equations.

In addition to numerical methods, advanced simulation techniques such as Monte Carlo simulations and quantum transport simulations are often used to capture the quantum mechanical effects and statistical behavior of carriers in nanostructured devices. These techniques provide a more accurate representation of the carrier dynamics and allow for the assessment of device performance under different operating conditions.

Furthermore, the integration of simulation tools with optimization algorithms and machine learning techniques has become increasingly important in the design and optimization of nanostructured LEDs and lasers. These combined approaches enable the exploration of large design parameter spaces and facilitate the discovery of optimal device structures for specific applications.

Conclusion

In this chapter, we have explored the simulation techniques used to model nanostructured LEDs and lasers. By employing rate equations and optical models, the electro-optical behavior of nanostructured LEDs can be accurately described and analyzed. Meanwhile, the modeling of nanostructured lasers requires more comprehensive models that incorporate coupled rate, propagation, and gain equations. The use of advanced simulation techniques, including numerical methods, Monte Carlo simulations, and quantum transport simulations, allows researchers to capture the complex physics and optimize the performance of these devices. As the field continues to evolve, the integration of simulation tools with optimization algorithms and machine learning techniques holds great promise for the design of even more efficient and advanced nanostructured LEDs and lasers.

Python Code Snippet

Below is a Python code snippet that implements the important equations and algorithms discussed in this chapter for modeling nanostructured LEDs and lasers.

```
import numpy as np
import matplotlib.pyplot as plt
```

```python
def simulate_LED(rate_equations, initial_conditions, time_span, dt):
    '''
    Simulate the dynamics of a nanostructured LED using the rate
    ↪  equations.

    :param rate_equations: Function representing the rate equations.
    :param initial_conditions: Initial conditions for carrier
    ↪  densities and photon density.
    :param time_span: The total time for the simulation.
    :param dt: Time step for the simulation.

    :return: Time array and state array containing carrier densities
    ↪  and photon density over time.
    '''
    num_steps = int(time_span / dt)
    time = np.linspace(0, time_span, num_steps)
    state = np.zeros((num_steps, len(initial_conditions)))
    state[0, :] = initial_conditions

    for i in range(1, num_steps):
        k1 = rate_equations(state[i-1, :])
        k2 = rate_equations(state[i-1, :] + 0.5 * dt * k1)
        k3 = rate_equations(state[i-1, :] + 0.5 * dt * k2)
        k4 = rate_equations(state[i-1, :] + dt * k3)
        state[i, :] = state[i-1, :] + (dt / 6) * (k1 + 2 * k2 + 2 *
        ↪  k3 + k4)

    return time, state

def led_rate_equations(y):
    '''
    Define the system of rate equations for the LED.

    :param y: Current state vector containing hole density, electron
    ↪  density, and photon density.

    :return: Derivative of the state vector.
    '''
    N_p, N_n, S = y
    G = 1e24          # Generation rate (example value)
    tau_p = 1e-9      # Hole lifetime (in seconds)
    tau_n = 1e-9      # Electron lifetime (in seconds)
    tau_s = 1e-9      # Photon lifetime (in seconds)
    R_sp = 1e20       # Radiative recombination rate (example value)
    R_sp_nonrad = 1e20  # Non-radiative recombination rate (example
    ↪  value)

    dN_p_dt = G - N_p / tau_p - N_p / tau_s
    dN_n_dt = G - N_n / tau_n - N_n / tau_s
    dS_dt = N_p / tau_p - S / tau_s + R_sp - R_sp_nonrad

    return np.array([dN_p_dt, dN_n_dt, dS_dt])
```

```python
# Simulation parameters
initial_conditions = [1e18, 1e18, 1e10]  # [N_p, N_n, S]
time_span = 5e-6  # Total simulation time in seconds
dt = 1e-9  # Time step in seconds

# Run simulation for the LED
time, state = simulate_LED(led_rate_equations, initial_conditions,
    time_span, dt)

# Plot results
plt.figure(figsize=(12, 6))
plt.subplot(3, 1, 1)
plt.plot(time, state[:, 0], label='Hole Density (N_p)')
plt.ylabel('Density (m^-3)')
plt.legend()

plt.subplot(3, 1, 2)
plt.plot(time, state[:, 1], label='Electron Density (N_n)')
plt.ylabel('Density (m^-3)')
plt.legend()

plt.subplot(3, 1, 3)
plt.plot(time, state[:, 2], label='Photon Density (S)')
plt.xlabel('Time (s)')
plt.ylabel('Density (m^-3)')
plt.legend()

plt.tight_layout()
plt.show()

# Note: Similar implementation can be done for laser rate equations
#    and simulations.
```

This code defines two functions:

- `simulate_LED` runs the time evolution of the LED carrier and photon densities based on a set of input parameters and initial conditions, using a numerical method to solve ordinary differential equations.
- `led_rate_equations` specifies the rate equations for the LED, accounting for generation, recombination, and photon density dynamics.

The simulation produces results showing how the hole density, electron density, and photon density evolve over time, allowing for the visualization of the LED's behavior under the defined conditions.

Chapter 31

Functionalization of Nanostructures

Introduction

In this chapter, we explore the techniques for chemical and physical functionalization of nanostructures to enhance semiconductor properties. Functionalizing nanostructures involves modifying their surface properties through the attachment of various organic and inorganic species. These modifications can significantly impact the electronic, optical, and chemical properties of nanostructured semiconductors, leading to improved device performance and novel applications. We delve into the mathematical and computational approaches used to understand and model the effects of functionalization on semiconductor nanostructures.

Surface Functionalization Techniques

Surface functionalization of nanostructures can be achieved using various techniques, including chemical functionalization, physical deposition, and self-assembly processes. Chemical functionalization involves the attachment of organic and inorganic molecules to the nanostructure surfaces through covalent or non-covalent bonding. Physical deposition techniques, such as physical vapor deposition (PVD) and chemical vapor deposition (CVD), enable the controlled deposition of functional materials onto the nanostructure surface. Self-assembly processes, such as Langmuir-Blodgett deposition and electrostatic adsorption, offer a bottom-up approach to arrange functional molecules in a well-defined manner on the nanostructure surfaces. These techniques provide researchers with a wide range of options to tailor the properties of nanostructures for specific applications.

Modeling Surface Functionalization

Mathematical models play a crucial role in understanding and predicting the effects of surface functionalization on the properties of nanostructured semiconductors. These models incorporate physical and chemical principles to describe the interactions between functional species and the nanostructure surface.

1 Surface Energy Model

A widely used model for describing the adsorption of functional molecules onto the nanostructure surface is the surface energy model. This model considers the competition between the surface energies of the bare nanostructure and the adsorbed molecules. The change in surface energy due to functionalization can be expressed as:

$$\Delta E_{\text{surf}} = \gamma_{\text{molecule}} A_{\text{surface}} - \gamma_{\text{bare}} A_{\text{surface}}$$

where ΔE_{surf} is the change in surface energy, γ_{molecule} and γ_{bare} are the surface energies of the functionalized and bare nanostructures, respectively, and A_{surface} is the surface area of the nanostructure. By minimizing the total energy of the system, the equilibrium configuration and coverage of the functional molecules on the nanostructure surface can be determined.

2 Quantum Mechanical Modeling

To capture the electronic and optical properties of functionalized nanostructures, quantum mechanical modeling techniques are often employed. Density functional theory (DFT) and time-dependent density functional theory (TDDFT) provide a theoretical framework to calculate the electronic structure and optical properties of functionalized semiconductors. These methods solve the Schrödinger equation self-consistently, accounting for the interactions between electrons and functional species. By simulating the absorption and emission spectra, the impact of surface functionalization on optoelectronic properties can be analyzed.

3 Kinetic Monte Carlo Simulations

In systems with a large number of atoms or molecules, such as functionalized nanostructures, kinetic Monte Carlo (kMC) simulations offer a powerful tool to investigate the dynamics of surface reactions and the evolution of functionalized structures over time. kMC simulations use random sampling to simulate surface processes on a lattice, incorporating the rates of adsorption, desorption, diffusion, and reaction of functional species. By running numerous simulation iterations, the time evolution of surface coverage and the distribution of functional species on the nanostructure can be obtained.

Applications of Surface Functionalization

Functionalizing nanostructures can significantly enhance their properties, leading to various applications in fields such as optoelectronics, catalysis, sensors, and biomedicine. By controlling the surface chemistry, the band gap of the nanostructured semiconductors can be tuned, enabling the design of new photovoltaic devices with improved efficiency. Surface functionalization can also enhance the catalytic activity of nanostructures, enabling more efficient chemical reactions. In the field of sensors, functionalized nanostructures can be employed to detect specific molecules or ions with high sensitivity and selectivity. Additionally, in biomedicine, functionalized nanostructures can be used as drug delivery systems or imaging agents, providing targeted therapies and diagnostics.

Conclusion

In this chapter, we have explored the techniques and modeling approaches used for the chemical and physical functionalization of nanostructures to enhance semiconductor properties. The ability to modify the surface properties of nanostructures allows for the tailoring of their electronic, optical, and chemical characteristics. Mathematical models incorporating surface energy, quantum mechanical calculations, and kinetic Monte Carlo simulations play crucial roles in understanding and predicting the effects of functionalization on semiconductor nanostructures. With precise control over surface functionalization, nanostructures can be employed in various applications, including optoelectronics, catalysis, sensors, and biomedicine. Through continued research and development, functionalized nanostructures have the potential to revolutionize several technological and scientific fields.

Python Code Snippet

Below is a Python code snippet that implements the equations and algorithms described in this chapter regarding surface functionalization and modeling nanostructured semiconductors.

```python
import numpy as np
from scipy.optimize import minimize
import matplotlib.pyplot as plt

def surface_energy_model(gamma_molecule, gamma_bare, A_surface):
    '''
    Calculate the change in surface energy due to functionalization.
    :param gamma_molecule: Surface energy of the functionalized
        nanostructure.
    :param gamma_bare: Surface energy of the bare nanostructure.
    :param A_surface: Surface area of the nanostructure.
    :return: Change in surface energy.
    '''
    delta_E_surf = (gamma_molecule - gamma_bare) * A_surface
    return delta_E_surf


def quantum_mechanical_model(func_species, num_orbitals):
    '''
    Mock function to simulate quantum mechanical properties using
        simple DFT-like calculations.
    :param func_species: Matrix of functional species properties.
    :param num_orbitals: Number of orbitals considered in the model.
    :return: Calculated electronic structure (for illustrative
        purposes, returns an array).
    '''
    electronic_structure =
        np.linalg.eigvals(func_species[:num_orbitals,
        :num_orbitals])
    return electronic_structure


def kMC_simulation(num_steps, adsorption_rate, desorption_rate,
    initial_coverage):
```

```python
    '''
    Run a simple Kinetic Monte Carlo simulation for surface coverage
    ↪  dynamics.
    :param num_steps: Total number of simulation steps.
    :param adsorption_rate: Rate of adsorption.
    :param desorption_rate: Rate of desorption.
    :param initial_coverage: Initial surface coverage.
    :return: Array of surface coverages over time.
    '''
    coverage = initial_coverage
    coverages = [coverage]

    for step in range(num_steps):
        if np.random.rand() < adsorption_rate:  # Simulate
        ↪  adsorption
            coverage += 1  # Increase coverage
        if np.random.rand() < desorption_rate:  # Simulate
        ↪  desorption
            coverage -= 1  # Decrease coverage
        coverages.append(max(0, coverage))  # Ensure coverage is
        ↪  non-negative

    return coverages


# Example parameters for calculations
gamma_bare = 1.5  # Surface energy of the bare nanostructure (in
↪  J/m^2)
gamma_molecule = 2.0  # Surface energy of the functionalized
↪  nanostructure (in J/m^2)
A_surface = 0.02  # Surface area of the nanostructure (in m^2)
adsorption_rate = 0.01  # Example adsorption rate
desorption_rate = 0.005  # Example desorption rate
initial_coverage = 0.1  # Initial coverage on the surface
num_steps = 100  # Number of steps for simulation
num_orbitals = 3  # Number of orbitals for quantum model

# Calculations
delta_E_surf = surface_energy_model(gamma_molecule, gamma_bare,
↪  A_surface)
func_species = np.random.rand(5, 5)  # Random matrix to simulate
↪  properties
electronic_structure = quantum_mechanical_model(func_species,
↪  num_orbitals)
coverages = kMC_simulation(num_steps, adsorption_rate,
↪  desorption_rate, initial_coverage)

# Output results
print("Change in Surface Energy:", delta_E_surf, "J")
print("Calculated Electronic Structure (Eigenvalues):",
↪  electronic_structure)

# Plotting surface coverage over time
```

```
plt.plot(coverages)
plt.title('Kinetic Monte Carlo Simulation of Surface Coverage')
plt.xlabel('Steps')
plt.ylabel('Surface Coverage')
plt.show()
```

This code defines three functions:

- `surface_energy_model` calculates the change in surface energy due to functionalization based on the surface energies and area.
- `quantum_mechanical_model` simulates the electronic structure using a mock approach resembling DFT calculations.
- `kMC_simulation` runs a simple kinetic Monte Carlo simulation to model surface coverage dynamics over time.

The provided example performs calculations for changes in surface energy, simulates quantum properties, and illustrates surface coverage changes over time, followed by plotting the results.

Chapter 32

Nanostructured Sensor Devices

Introduction

The field of nanotechnology has led to significant advancements in sensor technologies, enabling the development of highly sensitive and selective sensors for various applications. In this chapter, we delve into the advanced modeling techniques used to enhance the sensitivity and selectivity of nanostructured sensor devices. We explore the mathematical and computational approaches that enable researchers to optimize sensor design, understand the underlying physical principles, and accurately predict sensor performance.

Modeling Sensitivity in Nanostructured Sensors

The sensitivity of a sensor refers to its ability to detect and quantify small changes in a target analyte or stimulus. In nanostructured sensors, sensitivity can be enhanced through the manipulation of surface properties and the incorporation of specific materials. Mathematical models play a crucial role in understanding and predicting the sensitivity of nanostructured sensors.

1 Sensing Mechanisms

Different sensing mechanisms are employed in nanostructured sensors, including resistive, capacitive, optical, and piezoelectric effects. For example, in a resistive sensor, the interaction between the target analyte and the nanostructure surface leads to a change in electrical resistance. Modeling these sensing mechanisms involves considering the relevant transport phenomena and the interaction of the analyte with the nanostructure surface.

2 Surface Functionalization

Surface functionalization plays a pivotal role in enhancing sensitivity in nanostructured sensors. By carefully selecting the functional molecules or nanoparticles attached to the nanostructure surface, selectivity towards specific analytes can be achieved. Modeling the effects of surface functionalization involves understanding the adsorption dynamics, the interactions between the functional species and the analyte, and the subsequent changes in the sensor response.

3 Mathematical Models

Mathematical models are essential for simulating the response of nanostructured sensors to different analyte concentrations. These models often involve partial differential equations that describe the relevant physical processes, such as mass transport, charge transport, or light propagation. The governing equations are coupled with appropriate boundary conditions and initial conditions to capture the behavior of the sensor.

Modeling Selectivity in Nanostructured Sensors

Selectivity refers to the ability of a sensor to differentiate between multiple analytes, particularly when they share similar physicochemical properties. Achieving selectivity in nanostructured sensors poses significant challenges, but advanced modeling techniques can help in designing sensors with high selectivity.

1 Multivariate Analysis

Multivariate analysis techniques, such as principal component analysis (PCA) and partial least squares (PLS), are often employed to analyze the response patterns of nanostructured sensors to different analytes. These methods reduce the dimensionality of the sensor response data and identify the key features that discriminate between analytes. The resulting models can then be used to predict the concentration or identity of unknown analytes based on their sensor response.

2 Pattern Recognition Algorithms

Pattern recognition algorithms, including artificial neural networks, support vector machines, and decision trees, offer powerful tools for achieving selectivity in nanostructured sensors. These algorithms learn the relationships between the sensor response patterns and the analyte concentrations by training on a set of known samples. Once trained, the models can accurately classify unknown samples based on their sensor responses, enabling selective sensing.

3 Data Fusion

Data fusion techniques combine the output from multiple sensors to improve selectivity. By integrating the responses from multiple nanostructured sensors with different sensing mechanisms or surface functionalizations, a more robust and selective sensing platform can be achieved. Mathematical models that combine the individual sensor responses, such as weighted averaging or Bayesian inference, are used to obtain the final analyte concentration estimate.

Conclusion

In this chapter, we have explored advanced modeling techniques for enhancing sensitivity and selectivity in nanostructured sensor devices. Mathematical models play a crucial role in understanding the underlying physical principles, optimizing sensor design, and predicting sensor performance. By accurately modeling the sensing mechanisms, surface functionalization, and analyte interactions, researchers can design sensors with superior sensitivity and selectivity. The integration of multivariate analysis, pattern recognition algorithms, and data fusion techniques further enhances the selectivity of nanostructured sensors. Through continued research and development, nanostructured sensor devices have the potential to revolutionize various fields, including environmental monitoring, medical diagnostics, and food safety.

Python Code Snippet

Below is a Python code snippet that models the sensitivity and selectivity of nanostructured sensors using basic mathematical principles and machine learning techniques.

```python
import numpy as np
from sklearn.decomposition import PCA
from sklearn.svm import SVC
from sklearn.model_selection import train_test_split
from sklearn.metrics import classification_report
import matplotlib.pyplot as plt

def calculate_sensor_response(analyte_concentration,
    sensitivity_coefficient):
    '''
    Calculate the sensor response based on analyte concentration and
        sensitivity coefficient.
    :param analyte_concentration: Concentration of the target
        analyte.
    :param sensitivity_coefficient: Sensitivity coefficient of the
        sensor.
    :return: Sensor response.
    '''
    return sensitivity_coefficient * analyte_concentration

def perform_pca(data):
    '''
    Perform Principal Component Analysis to reduce dimensionality of
        sensor responses.
    :param data: 2D array where each row represents a sample
        response.
    :return: PCA transformed data.
    '''
    pca = PCA(n_components=2)
    transformed_data = pca.fit_transform(data)
    return transformed_data

def train_classifier(X_train, y_train):
    '''
    Train a Support Vector Classifier for selectivity
        identification.
```

```python
    :param X_train: Training data.
    :param y_train: Training labels.
    :return: Trained classifier.
    '''
    classifier = SVC(kernel='linear')
    classifier.fit(X_train, y_train)
    return classifier

def evaluate_classifier(classifier, X_test, y_test):
    '''
    Evaluate the classifier's performance using test data.
    :param classifier: Trained classifier.
    :param X_test: Test data.
    :param y_test: True labels for the test data.
    :return: Classification report of the classifier.
    '''
    predictions = classifier.predict(X_test)
    report = classification_report(y_test, predictions)
    return report


# Simulated data for nanostructured sensor response
analyte_concentrations = np.linspace(0, 10, 100)  # Simulated
↪ concentrations
sensitivity_coefficient = 1.5  # Sensitivity coefficient of the
↪ sensor
sensor_responses = calculate_sensor_response(analyte_concentrations,
↪ sensitivity_coefficient)

# Plotting the sensor response
plt.figure(figsize=(8, 5))
plt.plot(analyte_concentrations, sensor_responses, label='Sensor
↪ Response', color='b')
plt.xlabel('Analyte Concentration')
plt.ylabel('Sensor Response')
plt.title('Sensor Response Curve')
plt.legend()
plt.grid()
plt.show()

# Simulated dataset for selectivity
# Here, we assume each row is a response from a different sensor to
↪ varying concentrations of two different analytes
X = np.random.rand(100, 5)  # 100 samples, 5 features (responses)
y = np.random.choice([0, 1], size=(100,))  # Binary labels for two
↪ classes

# Splitting the dataset
X_train, X_test, y_train, y_test = train_test_split(X, y,
↪ test_size=0.2, random_state=42)

# PCA for dimensionality reduction
X_train_pca = perform_pca(X_train)
```

```
X_test_pca = perform_pca(X_test)

# Training the classifier
classifier = train_classifier(X_train_pca, y_train)

# Evaluating the classifier
report = evaluate_classifier(classifier, X_test_pca, y_test)
print(report)
```

This code defines several functions:

- `calculate_sensor_response` computes the sensor response based on analyte concentration and sensitivity coefficient.
- `perform_pca` executes Principal Component Analysis to reduce the dimensionality of sensor response data.
- `train_classifier` trains a Support Vector Classifier for identifying selectivity based on sensor response.
- `evaluate_classifier` evaluates the trained classifier's performance against a test dataset and outputs a classification report.

The example generates simulated sensor responses, visualizes them, and demonstrates how to train and evaluate a classifier for selectivity identification based on the sensor responses.

www.ingramcontent.com/pod-product-compliance
Lightning Source LLC
Chambersburg PA
CBHW052150220526
45471CB00004B/1615